Locally Laid

Locally Laid

How We Built a Plucky,
Industry-Changing
Egg Farm—from Scratch

Lucie B. Amundsen

AVERY
an imprint of Penguin Random House
New York

AVERY

an imprint of Penguin Random House LLC
375 Hudson Street
New York, New York 10014

First trade paperback edition, 2017

Most Avery books are available at special quantity discounts for bulk purchase
for sales promotions, premiums, fund-raising, and educational needs. Special
books or book excerpts also can be created to fit specific needs. For details,
write SpecialMarkets@penguinrandomhouse.com.

ISBN 9780399185601 (paperback)
ISBN 9780698404052 (eBook)

Printed in the United States of America
3   5   7   9   10   8   6   4   2

Book design by Ellen Cipriano

Some of the names and identifying characteristics have been changed
to protect the privacy of the individuals involved.

To Jason, Abbie, and Milo
for giving me
a storied life

All photos were taken at Locally Laid Egg Farm in Wrenshall, Minnesota.

Contents

Act 1: Hatch

Act 2: Cluck!

Act 3: Egged

ACT 1

Hatch

VERB: to emerge from incubation;
to conspire to devise a plot

Chapter 1

2012

At dusk, hens seek their coop. So reliable is this, there's even a saying, an adage: *Chickens come home to roost.* It's for warmth. It's for protection. It's hardwired. But our first shipment of nine hundred mature birds, just purchased from a commercial operation, stands on the field staring. They tilt and turn their heads to better align us with their side-placed eyes, as though awaiting instructions.

Then, as darkness quiets the pasture, I get it.

My hand on my lips, I mumble, "Oh, God."

These hens are out of sync with sunset because until today, they have NEVER SEEN THE SUN. While I've worried about many things going wrong with our unlikely egg startup, CHICKENS not knowing HOW TO BE CHICKENS was not one of them.

2010

Two summers before I'd ever heard the term *pasture-raised*, handled a mature hen, or imagined our egg cartons in grocery stores, my husband and I were enjoying a rare weekend alone. Our school-aged children were trooped off to his mother's town house in the Twin Cities. I can't remember what our children were doing in Minneapolis that weekend, but likely they spent time in their grandmother's community pool. The weather was perfect for it.

Even up in northern Minnesota it was hot, hotter than we've come to expect for this not-quite-summer month in Duluth. Spring is slow to come here. For perspective, my lilac bushes don't pop a blossom until midway through June. We're still scraping our cars while friends' flower beds just hours south in the Twin Cities look like a Georgia O'Keeffe painting in heat.

But this particular June day in the Northland, I wanted nothing more than to be outside. I set about tucking fragrant petunias and tall grasses into the large planter at our home's front door, while Jason made another go at his chicken coop project, behind the garage. I couldn't see him, but the banging and swearing wafted over.

One could say this was my fault.

Late the summer before, I'd written a story about Duluth's new poultry-friendly city ordinance. Lots of communities have something like this now. In fact, five hens allowed in town is so commonplace it's practically a non-story, but in 2010 Duluth was an early adopter in the urban farm movement. And Jason was unexpectedly spun up in its "grow your own food" zeitgeist. Until then he'd shown little interest in keeping a vegetable garden, or even in lawn care. But after my article hit newsstands, he'd piloted our Toyota minivan to a flea market just over the bridge to Wisconsin. He'd returned with several sheets of corrugated metal and a good-sized domed skylight strapped to the rack. The window had been rejected from a Walmart construction site for a hairline crack; Jason felt it was nothing a little duct tape couldn't handle.

The following spring, with at least one complete teardown and do-over, the result was a home chicken coop that looked a bit like a handicap-accessible Porta-Potty. And though he'd had significant help from our neighboring carpenter, Matt, it wasn't a bad attempt for a man who'd never so much as built a bird feeder.

Jason had been actively working on it since early spring and though it was not yet complete, we'd long had chicks. Seeing the fluffy baby birds at the feed store that early spring, he hadn't been able to restrain himself, coming home with five little peepers—their coop home barely even begun. During the construction period, they lived in our garage.

Like all our pets, the chicks immediately bonded with Jason, scrambling to him in our urban yard when they heard his voice. Soon they'd outgrown their kiddie pool home where they'd chirped nonstop under a heat lamp, all fluff and peep. Now they were deep into their awkward adolescence: beaks growing faster than faces, feathers sprouting over down, and legs rapidly outstretching coordination. It's a universally cruel stage.

The hens, now uncontainable tweens, freely roamed our garage at night, defecating indiscriminately. I tried to be a good sport as I wiped their brown-and-white poo dollops from my Christmas storage bins, deciding that seeing Jason buoyed by these birds was worth a certain amount of shit on a box.

The chickens spent their days in our mostly fenced-in yard, walking about with their jaunty, robotic manner while absurdly chattering on—*CO-KE . . . coke coke coke coke.* They murmured pleasantly, scratching the earth hunting for bugs and seeds while Jason stood by, beaming like he'd hatched them himself. To be fair, I also felt a sense of secondhand joy as I watched a hen open her wings forward, beak into the air, and stretch back a single leg into what we called the chicken yoga pose. That's a happy bird.

And I saw that these chicks meant more than just the promise of fresh eggs. For Jason, it was a step toward modest self-sufficiency. While we both come from a line of people who can damn near fix anything, we cannot. I tease that we're remedial adults and while I don't like that feeling of helplessness, Jason was actually doing something about his deficits. He was working the poultry angle as an independent study toward his life skills GED. Overall, it'd been good. This self-directed project, worthy of a 4-H ribbon, had renewed his confidence and brought him hands-on satisfaction, something that his office life managing grants at a politically charged hospital behemoth did not.

Just the week before, while digging beetles out of the garden to hand-feed "the ladies," he'd prattled on about how things "were really turning around for us." In some ways, he was right. We'd recently bought a house with a modest view of Lake Superior and moved out of an awkward rental situation. And because our new house has a little mother-in-law apartment, we'd rented out to a college student, and our cash flow was trending positively.

But, to be fair, Jason often feels that happy pull of life's pivotal turn. For him, we're always a heartbeat away from that elusive place where everything will come together.

"It's all right around the corner," he assured.

I nodded. Despite my suspicions that we actually existed in a round room.

On my knees at the planter, a shadow cast over me.

"Lu," Jason said, and I looked up. Dirty in his red ball cap and leather gloves, he was charming, like a boy playing farmer. "We should go out for Mexican. Just me and my Bird."

This delighted me.

Married for almost a decade, I love that he still calls me that.

Nearly every boyfriend I'd ever had has called me Bird. The consensus is accidental, but no doubt the endearment is appearance based. Not only did I inherit my father's nose, a thoroughly unapologetic appendage, I got it to exact scale.

There's an upswing, though: my "nose advantage." As surely as it has protected me from shallow men, it's also made me work harder, foster a sense of humor, and, perhaps, be a little kinder. It was obvious at a young age; I wasn't going to make it on my unique beauty alone.

So when I hear the tenderness with which Jason calls me

Bird, it feels like a little love song from someone who's passed a test of character. It's my silver lining to a life bereft of the normal use and enjoyment of champagne flutes and shot glasses.

I smiled. We were going on a date.

Showered and changed into one of my favorite summer flirty skirts, I enjoyed the happy buzz from the day's sun, the cold beer, and the warm chips. Even the loud chatter of Mexico Lindo patrons, which can sometimes jostle my nerves, seemed pleasant that night. It was a warm gift of an evening after a beautifully productive day. And that was when Jason said what every girl out with her fella longs to hear.

"I want to talk to you about something," he said, clearing his throat. "Commercial egg farming."

If this were a sitcom, a record needle would scratch across vinyl and someone would cue the laugh track. But as this was just my life, I blinked and kept shoveling salsa into my mouth between gulps of beer. Buzzed on Corona and that special hopeful feeling a woman gets about her man, I dodged.

"Yeah, well, I really like the chicks, too," I said, not looking up. "Mmm . . . I gotta stop with the chips, but this salsa is so fresh."

"Lu," Jason pushed on, "I've read these books on raising chickens on pasture, Joel Salatin books . . ."

I vaguely recalled the name. Wasn't he in *Omnivore's Dilemma* or maybe one of those disturbing food movies?

"Chickens on pasture . . . like free-range eggs?" I offered.

Instinctually, I knew he wasn't talking about whatever cage-free eggs are. While I wasn't then sure what the exact distinction entailed, I'd been a writer long enough to sense something too careful in the phrasing "cage-free" to convince me those birds were out frolicking in the fresh air.

"No, this is much better than that," he said with a hint of annoyance. He explained that free-range chickens (using cheeky air quotes to emphasize the term) often go out into the same field day after day until it's pecked down to nothing but dirt, hardpan. That is, if they actually go out at all. It's one of those terms that's been co-opted by marketers and can now mean a warehouse packed with tens of thousands of birds and just a few small doors leading to an outdoor concrete patio.

"Pasture-raised birds," he said proudly, "are rotated onto fresh grass every few days, so the field is actually part of their diet. It reduces food costs, plus the hens get exercise and so are, you know, better for it. Their eggs, too."

Jason prattled on about how some 95 percent of America's chickens are in battery cages—that's a poultry housing system made up of neatly lined wire crates. With several chickens in each cage, it allows for hundreds of thousands of birds stacked in a single building.

Then he moved on to the subject of cage-free birds. While an improvement, these chickens wander freely around their ware-house, but with no access to the outside world or even natural light.

"They're not in cages, but they can never leave—so it's like one big cage."

"A little like Hotel California?" I offered, but Jason did not acknowledge my vintage Eagles joke.

That left the category of pasture-raised chickens. It seems they're living the poultry dream—and, according to Jason, we could be, too.

I nodded as I took another swallow of beer. I didn't say that it sounded like an enormous amount of work or that we live in arguably one of the harshest climates in the continental United States. Nor did I point out that having spent our entire careers jockeying keyboards to make a living, we are not farmers.

So while I didn't exactly tune him out, I became a passive listener. A very passive listener. Poultry wasn't exactly the foreplay talk I was hoping for, so instead I just enjoyed the rhythm and cadence of his voice. I heard something about pastured hens foraging on fresh grasses producing healthier, delicious eggs with less fat and cholesterol, something about the local food movement and its ability to remake America's food system.

I signaled the server for a second beer and let it all wash over me with an occasional nod until an utterly un-ignorable statement pulled me out.

"This is the kind of farm I want to start," he said.

Now I was listening. In fact, I was listening so hard I realized that this particular corner of the restaurant was a convergence point for the piped-in music from two separate rooms, and they were competing against each other like dueling mariachi bands. Across from me, Jason was searching my face for traces of excitement about his decision to answer the noble poultry calling.

Start a farm? I thought. This is a man who until a few years ago could not identify a pear.

I swallowed, understanding I would have to steer hard and fight inertia to keep this date on track.

"Jay, honey, let's just be happy right here, right now. I mean, we're living in Duluth like you wanted. We've finally got real health insurance, and the kids have adjusted well."

"Lucie," he said with clear exasperation, "this is so much bigger than that."

Had I not been listening when he used "Itsy Bitsy Spider"–like hand gestures illustrating the sun nourishing the grass, which shelters the breeding insects that augments the pasture diet of the grazing and exercising birds, making for superior eggs?

I got irked and defensive. Not that he believed I hadn't followed the ecological circle of life he'd demonstrated, but rather defensive of the life we'd built.

"Listen, Jason, for like two goddamn minutes can't we just BE without chasing some . . . elusive something else?"

This was louder than I intended and the couple sitting next to us eyed me. I gave them the "everything is under control here" sheepish and apologetic smile-nod, and I felt our beautiful evening bank and career off course.

I cleared my throat, took a breath, found a gentle smile somewhere in my increasingly tense body.

"You have chickens now and the coop is coming along great," I offered as kindly as I can. "Why not do that for a while? See if you like it?"

"No, this is more than a backyard flock; you don't understand how this could change our lives." Jason's gaze was now intense.

Oh, but I did understand.

I understood that we'd make big-ass fools of ourselves and probably lose everything in the process. Jason leaping into agriculture with a poli sci degree, a master's in international affairs, and five pubescent chickens shitting in our townie garage is the working definition of asinine. The word's got *ass* right in it.

"All right"—my thoughts scrambled as I became aware of my breathing—"but you're not saying you're going to leave your job to be a chicken farmer, right?" I said it with forced lightness, a softball statement meant to be disputed.

"Well, I wouldn't quit right away—but yeah," he said. "I want to build a local egg-laying business . . . with cattle, too."

No hedging, no sugar-coated vacillation. There it lay, like a turd on the table between us. Hopeful mirth packed her little bags and left the room. My beer glass came down too hard and the concerned couple next to us looked up again. But I was done with niceties. Soon, they were flagging the server for the check.

"Lu, it's a perfect system," he said, continuing on about people wanting to feel connected to their food, as I crossed my arms and legs and adjusted my weight in an annoyed cant at the back of the chair. As he talked, all I could hear was the rush of blood through my ears. I was still stuck on cows.

My mind flashed to the ag majors in my dorm at the University of Maine, a land-grant university outside Bangor. They wore tall muddy boots and had long rubber gloves that went past their elbows for exploring a cow's nether regions. Jason didn't own gloves like that. And something about the concreteness of that fact fueled my outburst.

"Jason, you are NOT a farmer. You're a guy who had an idyllic youth in a wealthy suburb," I said, leaning in with a sharp edge. "You have pushed me on too many major life changes to pursue ambitions—YOUR ambitions." I thought of how I've contorted to accommodate his varying and ever-changing pursuits.

My voice pushed higher as I went on.

"I moved to this godforsaken Arctic for you. Because you told me . . . you told me it would make you HAPPY!"

On that last part, I thought I might reach across the table and snap Jason's head off.

We're not a couple who fights and certainly not in public. And usually, if I push back on something, Jason can be counted on to put his crazy notion du jour into a holding pattern, to think it through a little more and, over time, see reason to abandon it. Ideas that have died this way include but are not limited to the following: opening a wind turbine collective, starting a hedge fund, and buying a sailboat to transport goods via Lake Superior. (The fact that I'm a rail-hanging vomiter on open water probably helped nix that one.)

It's always been less about actually pursuing all these projects and more about his being an external processor, airing out all these wacky thoughts. Giving them a little wave on his mental flagpole. I also know that upsetting me distresses Jason. I am, after all, his Bird. And usually he comes to terms with the fact that he is a grant writer. And a pretty good one. So it was inconceivable when, instead of rolling up this unpleasant topic,

Jason leveled a glare at me and said slowly, "You cannot stop me from pursuing this."

I tasted the bitterness of adrenaline and betrayal in my constricted throat. Thoughts binged around my head in an enraged electrical storm. *How dare you cast me as the pisser-on-er of dreams!*

I put up a halfhearted fight against angry tears and succumbed.

Having done my fair share of food industry service, I can tell you that nothing, *nothing* unnerves waitstaff like the fighting couple with the crying woman seated in one's station. I remember glancing up for a second and seeing our waitress do a sort of horrified back-and-forth shuffle with our meals as she looked for an opportunity to throw them on our table and bolt. But that right moment was elusive, as I wasn't engaged in dignified weeping into one's napkin. I rarely cry, but when I do, it's full-on bawling with shallow, gasping breaths. The kind that can really get the snot flowing. The kind that produces honking.

To understand my seemingly ungenerous reaction, my refusal to even entertain this man's fervent dream to enter the commercial egg economy, one must pause the mariachi bands for just a moment to understand our past.

I left a good life in the city.

During my fifteen years in Minneapolis, I made the kind of friends who'd take a two a.m. phone call and who got me through the year of Jason's homeland military deployment when Abbie was a toddler and Milo was still tucked into a sling. I also created a career. As an editor for a *Reader's Digest* publication based in the Twin Cities, I wrote DIY articles teaching faucet installation, lawn maintenance, and how to build a closet organizer out of a sheet and a half of veneer plywood—in a weekend. It was functional, surprisingly lucrative work and, most importantly, made me feel useful.

We'd also settled in our forever house. The one I called "The Beige Rambler of My Dreams," a truly solid 1950s atomic rambler. Thick plaster walls, an open floor plan on a corner lot in an award-winning school district; it was a lovely home. Add the large windows, a couple of fireplaces, and a finished basement large enough to raise Shetland ponies, and, well, it was where we were going to raise our children and grow old. Now that we had poured two years of our lives into remodeling it and more money than I'd care to admit, it was nearly perfect.

Nearly perfect enough to be quietly killing Jason.

Don't let me mislead you. He'd wanted to purchase the place, even more than I did, as we pushed ourselves right up against our financial limits to make it happen. We both yearned for a good education for the kids, a quiet neighborhood feel, and all the trappings of the middle class that we had both enjoyed growing up.

But this first-ring suburb put Jason on the verge of a boredom aneurysm.

That was when he got the job offer to be a grant writer for a hospital on Lake Superior. They proposed we crowbar our family from the culturally rich and economically robust Twin Cities and move to a city annually ranked among the coldest locations in America with a job climate to match.

Duluth, Minnesota, was as much a punch line as a home to 86,000. And it couldn't care less. Situated by the world's largest freshwater lake, it has seven miles of white-sand beach, more urban hiking and biking trails than any metro location in the nation, and a streak of—as my friend Jake calls it—the Wild Midwest.

There are just fewer rules here. People are drawn to the untamed aspect of this place. Especially the kind of people who wished they lived bold lives rather than, say, typed grants.

When I agreed to move over two hours, two growing zones, and seemingly two planets north for the promise of good health insurance and a happier spouse, it wasn't easy. It was 2008 when we naively planted the For Sale sign in front of my dreamy rambler—the very same month that economists now say the housing bubble went pop.

It didn't sell.

Jason moved to Duluth without us to start the new grant-writing position at the reassuringly venerable hospital system. I was left with two small children and a big dog living in a real-estate-staged house, which isn't really living at all. After nearly seven months and forty-three house showings that produced no offers—not even an insulting lowball one—we started brain-storming ways to reunite. We came up with the idea of renting a place in Duluth, but it would have to be on the cheap.

After walking through many rough apartments, Jason was unexpectedly looking at a house within our rental budget. On the beach, no less. He called me while I was at PetSmart with the children, hiding from another fruitless real estate agent walk-through.

"It's a three-bedroom rambler! With first-floor laundry! And an attached garage!" he enthused, sounding the suburban house-wife mating call. "Clothesline! Fenced-in yard! Master bath!"

I was so excited I could hardly contain myself. The children, sensing my joyful discombobulation, started gesturing wildly for a gerbil—apparently, at this moment, I was a vulnerable adult.

"But," said Jason, "there's just a couple things."

I stiffened, held my breath, and the children knowingly retreated.

"It's right next to a church," he said.

"Yeah?"

"And there is a life-size statue of the Virgin Mary in the yard."

"Oh," I said. "Well . . . I'm down with Mary."

"And it's really close to the church because, well, it's actually the rectory."

"Oh," I managed. "That's . . . truly different."

"And the church has no running water"—this is where he started talking fast—"so . . . so per the lease agreement, on Sunday mornings from eight thirty to ten in the morning, parishioners can use the bathroom."

Which gives new weight to my thought: *Holy shit. The economic downturn is driving me into a semipublic restroom situation.*

I swallowed and heard myself say "That's okay," and made a mental note to get the really big container of Clorox wipes.

Within weeks, we'd rented our suburban Minneapolis home and moved into the rectory, complete with the wall-sized print of the Last Supper in the dining room. Certainly, our beautiful home would sell in a few months, and then we'd join the wild buyer's market that had real estate agents and home seekers completely lathered.

We lived at that rectory, with its big Virgin Mary and full-bladdered parishioners, for more than year and a half—a time during which my then five-year-old son, Milo, completely taken with our proximity to Lake Superior, listened to "The Wreck of the Edmund Fitzgerald" on repeat.

I tell you all this—the Beige Rambler, the rectory lifestyle, the months of single parenting—to handily prove my support for this man across the table sitting squarely in my temper's sight lines.

I'd moved; I'd shared my bathroom; I'd listened to Gordon Lightfoot.

Back at the restaurant, with its battling Mexican horns and guitars, two years have passed since the housing crash and the country still has not fully recovered. And though I am still free-lance writing and indeed file copy with nearly every publication in my new city, I'm hardly making anything. Somehow I fell from breadwinner to bread eater and honestly, there was significantly less of it these days.

My dim prospects in Duluth—a place where I could not land a job interview, much less a job—finally drove me to the ultimate home for wayward professionals: graduate school. I was slated to start at the end of the summer.

Oh, and this is where I have to set the record right on a couple of things.

When I yelled at Jason that we are always doing things to

realize his dreams and never mine? Well, that's not completely true. He was fully supportive of my returning to school for my master's, which is more impressive a commitment than it may sound. You see, Duluth doesn't have the program I'm seeking. What I want is a master of fine arts in writing—and yes, I realize that's like saying, "You know that thing I can't make money with anymore? More of that, please." But looking through Duluth's slim want ads, I'm convinced that writing is truly my only skill, as useless as it feels these days.

But with this particular kind of degree, which is a longer program and requires a book-length thesis, I could parlay my writing résumé into teaching at the postsecondary level. Because it's the terminal degree, the highest attainable in a given course of study, I could, one day, become tenured faculty and enjoy all the security that would provide. (My dreams were clearly less imaginative than Jason's tilting turbines and expansive poultry endeavors.)

The master's program is also terminal because it's three hundred miles round trip from Duluth, not online, and is bound to kill us all. That it's located ten minutes from the Beige Rambler tells me that I should study literary irony.

Lastly, that crack calling our new region a godforsaken Arctic? Well, that wasn't really fair to Duluth. While no doubt it's frigid, that enormous lake seems to add a needed ballast to life, slowing the pace enough so that kids get to be kids a bit longer. Or maybe we all get to be kids longer here. Back in the Twin Cities, I felt people were squarely defined by their jobs, but in Duluth employment is treated as what you do when you're not pursuing your passions—maybe your music or a visual art form or distance kayaking. I like to joke that one can't swing a dead seagull up here without hitting a poet.

But at that moment, sitting in Mexico Lindo, I don't remember how encouraging Jason was about graduate school or all the things I liked about our small city by the unsalted sea. I don't remember eating. And I don't even remember getting home. What I do recall was Jason sleeping on the couch and my going to bed in my clothes.

And one clear thought: We can never, ever go back to Mexico Lindo.

Chapter 2

J ust a day after the restaurant "farm-gument" with all its the-
atrical howls and hand gestures, I subsided into quiet, ratio-
nal conversation. I gently walked Jason through the many, many
reasons why we were not the people to undertake this egg venture. Worse than not being farmers, we were students of liberal
arts, I'd explained. We lacked the skill set for such a venture (or
it seemed in those days, most ventures). And it wasn't like we
could start a new business right then anyway. All our capital was
currently tied up in real estate, held hostage by the great housing recession. So, I'd concluded, as fine a cause as freeing chickens was, it was just not possible for us right now (or ever).

While it may be uncouth to assess a quarrel in such bald
terms: I was winning. In my appraisal over the past twenty-four
hours, I'd managed to repack a solid 89 percent of Pandora's box,
and honestly, I think we would have placed it on a high shelf had
not two events, unforeseen and ruinous, toppled our tit-for-tat
marriage economics.

It started just three days after our Saturday dinner out, and unlikely as it sounds, it continued its life-changing rampage all the way through Southeast Asia and back, irrevocably changing the poultry trajectory of our life path.

That Tuesday afternoon, our weekend quarrel more or less behind us, I was weeding our small garden when I heard the crunching of gravel in our alley.

"Hey," I said, walking, chives in my hand, to meet our minivan. "Nice surprise seeing you home early, honey."

Jason always changed at the hospital after his couple-mile walk to work, so it was rare for me to see him in his dress clothes. His sport-coat-and-loosened-tie handsomeness made me smile. It hadn't yet registered that it was unusual for Jason to have the vehicle at all.

Expressionless, he replied, "I wish it were," and, as Jason said it, he swung the van door open to reveal the fish tank from his office.

I'd furrowed my brow. *Why is this here?* It was like a child's riddle with an elusive answer. I mean, this was the tank that I'd banished to his work office when we moved up north. If the fish are at home, that means . . .

"Oh my God." As I put my hand to my lips, my eyes met his.

His crumpled look confirmed he'd rather do anything right now than tell me his position had been eliminated. Jason understood I'm his stability-craving gal and our foundation was about to be kicked out from under us. It had been this position's stolid reliability that crowbarred us out of my beloved Minneapolis life. It was the ballast keeping us on the right side of the middle-class line.

"I'm so sorry, Bird," he said, his voice scuffed with emotion.

"No," I said, shaking my head.

I turned away, crunching the rocks under my feet, my brain rejecting that his grant writing position was gone. "THIS job that came looking for us. And you have great performance reviews and . . . and . . . bring in several times your salary."

Jason reached out for my hand, but I jerked away.

"We MOVED here for them. And you sought their counsel before we bought this house," I said, flinging a hand toward the modest home behind me. "And they *encouraged* us to buy. Said we should absolutely invest in Duluth. Did you remind them of that?"

I suspect he answered, but the words bounced off the white static in my head. What I kept landing on was that this wasn't just wrong; it was illogical. And if I could simply pile up enough reasonable facts, it would set everything back where it was, where it should be. The fish swimming in his office, the tie straight around his collar, Jason tucked behind his hospital-issued PC.

"Do they understand what we've been through for this job? That our Minneapolis house is still unsold? That we lived in a rented rectory where parishioners used our bathroom? And"—I swallowed—"everything I gave up?"

It was thoughts of these last two years and my inability to find real work that broke me into tears. I stopped flailing long enough for Jason to tuck in my arms and pull me into his chest. The chives fell on my sandaled feet.

Over the next day, we oscillated between who was comforting whom, as one of us would take point, directly facing this untenable situation, while the other would draft behind, shouting out platitudes of encouragement. "It'll all work out! Everything happens for a reason!"

The reality was we still had an unsold house in the Cities and this new little home on Duluth's hillside. Our friend Paul playfully called us "land barons"—which we all know is fancy talk for scared multiple-mortgage holders. It wasn't too hard to imagine a scenario in which we lost everything. In fact, we seemed well on our way.

As unflattering as it is to admit, I was feeling sorry for myself. But had I known what was ahead, I would have employed the oft-said maxim of my mother-in-law, Mimi: "Be grateful for the problems you have."

The next day, we received a call that obscured all employment worries and set our lives in a slingshot pointed to Asia. Jason's younger brother, Brian, an expat living in Cambodia, had suffered a severe drug overdose and lay in a coma in Phnom Penh—half a planet away. And it was there that a medical team, impatient to clear the bed, was eager to pull life support.

This triggered a flurry of our own calls: to the State Department, the U.S. consulate, and two U.S. senators' offices to expedite Jason's lapsed passport.

In the three hours it took Jason to drive to the Twin Cities and meet up with Mimi, I was part of an online, telephone-attached team of relatives who secured travel papers and booked their flights.

The whole event sparkled with a breathless sort of frantic.

As Jason hastily exited the continent, he'd left a few loose ends. Like the chicken coop. While the mostly completed structure now housed our small urban flock of five birds, there was no fenced-in run yet. No place for them to exit their shelter via a small doggy door to a penned-in area of outdoors. What this meant was that every morning, the kids and I would have to open the coop door, scoop up birds (some willing, some not), and place them into a chicken tractor.

Okay, stop. That image in your head of a bird in the cab of a John Deere? Scratch that. A chicken tractor is nothing more than a portable box; ours was five feet long, two feet tall, and three feet wide, covered with chicken wire. It has no floor, for ground access. Its name has nothing to do with the structure, but rather the tractorlike function performed by the birds themselves. Inside, as chickens forage for grasses, bugs, and seeds, they dig, weed, and fertilize the land as they go—all tasks modern tractors perform.

While some chicken tractors are on wheels, ours got hand-lugged to different parts of our city lawn—which Jason endearingly called pasture. It was a labor-intensive way to do things.

Peeling back the metal flap on top of our chicken box, I was reliably greeted by much worried peeping. Over the past few months of hen stewardship, I'd grown confident in just reaching in, seizing a bird, and tucking her under my arm. She'd feel safe there and I'd avoid a wing cuff to the face.

My favorite chick was the tawny-colored Buff Orpington. She promised to one day be a bodacious plus-sized model of a chicken, wearing fluffy pantaloons under full feathery skirts and with as charming a personality as her appearance suggested. Predictably named Buffy, she didn't mind being handled and rather seemed to enjoy the company, clucking softly with a closed beak as I picked her up and stroked her silky feathers.

With my hand, I palmed one of her scaly feet. It looked more like a jumble of pointy sticks than an actual pedestrian appendage. In the front were three leathery toes, each tipped with a long talon, and a shorter toe rounded out the foot's back. They betrayed her dinosaurian genetics. I read once that chickens are a third cousin to Tyrannosaurus rex, more than 100 million years removed. I'm not surprised. There's a real ancientness to a creature that must turn its entire head to take you in with one side eye.

Transferring the birds from tractor to coop was uncomplicated, except for Skip—a glossy, black sex-link (a bird crossbred to associate a certain color with its gender, to make chick sexing easier). She was named after a friend who'd once proclaimed backyard chicken keeping "the stupidest idea ever." Poultry Skip was wily like her namesake and when I tried to grasp her, she'd dodge my snatches, unafraid to peck me. When I did manage a half purchase, she'd fidget out one of her wings and frantically flap it in my face until I was forced to unhand her. Then she'd bolt down the alleyway. This took what should have been a twenty-minute task of evening chicken chores—freshening up water, shaking more bagged chicken feed into their poultry trough—to an hour or more as the kids and I hunted among the trees and brush that lined the alley.

Chickens, it seemed to me, were a bit of a bother.

As the days piled into weeks, I took on additional sideline writing projects, like updating Jason's résumé, writing daily posts on the CaringBridge blog about Brian's health . . . and quietly typing out an obituary draft, just in case. And while always on the hunt for work for myself, I now applied for Jason, as well.

Writing letters and electronically forging his name, I buoyed myself knowing that at least the farm fervor had been ground out of my husband. When we talked on the phone, it was clear Jason was spent from too little sleep and too much responsibility, all the while navigating the unfamiliar and strange world of Cambodia. From my perspective, he was a raw dangling nerve, primed for the next available carpeted chute heading to a climate-controlled cubicle, any cubicle, adjacent to an office-sized Bunn coffeemaker. Jason ached for the familiar and this Asian hospital had none of it.

In that part of the world, family involvement in hospital care is crucial. While there are competent nurses, there are fewer of them and the patients who fare best have loved ones tending to them. Jason, one could say, grew into this role, while his mother was a natural as a trained nurse. He learned to change Brian's adult diaper and sheets, grind medicines with water to push into his nasogastric (NG) tube, which ran up his nostril and directly into his stomach, and was constantly advocating for more aggressive treatments.

Jason was recounting all this when I interrupted him before the phone card minutes ran out.

"Hey, I know this is a little off topic, but one of the chickens laid her first egg today," I said. Milo had found it in our backyard coop that morning.

"Her first egg? A pullet! She's early! I missed it! What color is it? What size? Who laid it?"

Jason seemed to oscillate between feeling a rush of poultry pride and bereft for not being there for this first experience.

I texted a picture of Milo holding out a light brown egg to the cell Jason had rented in Asia. Mimi would tell me later that he showed that photo to anyone whose English was nuanced enough to appreciate that this egg was from his own hen.

While I didn't relish picking up Skip the chicken any more than I did before, I had a sense of gratitude to the entire flock for providing Jason a small diversion, a blip of happiness in his otherwise medicalized world.

About three weeks in, the news out of Cambodia was a roller coaster of tantalizingly positive tidbits—*Brian was emerging from the coma! He was off his ventilator!*—followed by tragic updates of rollicking seizures that kept Brian's care team up twenty to thirty hours at a time. Eventually, his pupils went fixed and dilated—a symptom that often indicates injury to the upper brainstem and little hope of recovery.

It was then that a medical staff member, not on Brian's care team, floated an idea. What if Brian was given a small amount of the prescription drug he'd overdosed on? Brian's physicians forbade it.

This pissed Jason off. He was antsy, ready for action, and probably the fact that he'd recently been assaulted didn't help matters. But it may have, inadvertently, changed everything.

Just a few days before, Jason had been part of the noisy streetscape, trying to talk to his aunt Joyce back in Shakopee, Minnesota. To avoid the blaring traffic and techno music, he'd ducked into a quiet construction site, phone pressed against his ear, eyes on his shoes. That was when a hard punch connected with his cheekbone. The phone went flying.

Probably the worst text I've ever gotten was the one line, *Jason's been mugged.*

Recounting it later, he would say his military training must have kicked in.

"Before I could think about it, I'd kicked the legs out from under one of the guys."

And that was when he said it. Jason uttered a phrase so outrageous, so utterly shameless, it can be used only once per lifetime, and until then stored in a special box sternly labeled, *In case of emergency, break glass.*

"It's terrible; it's right out of a Steven Seagal direct-to-VHS movie," he admitted, as I coaxed the story out of him again. "Well, I mustered up my army drill sergeant voice and I barked, *'Motherfucker! You want a piece of me?'*"

Jason claims the second it came out of his mouth, he was

already embarrassed. Embarrassed in front of what turned out to be teen boys, kids really, who clearly didn't speak English. They ran off with his phone and Jason found his way back to Brian's hospital room with a headache, a purple contusion, and a strong will to get his brother well—and the hell out of Asia.

This fiery vigor motivated him to ring up one of his oldest friends, a medical doctor. They had a lengthy conversation about whether Brian should be administered the suggested, yet forbidden, medicine, and the phone call ended with Jason riding the back of a moto (the bastard offspring of a moped scooter and a BMX dirt bike) to a street pharmacy across town. There he bought the drug with cash.

And like he'd been doing for weeks, Jason ground up the pills, mixed it with water, and shoved it up Brian's NG tube. I can only guess that if he'd done this in an American hospital, he would've been arrested.

No one can say with certainty the effect of giving Brian those meds, but it was after he received them that his recovery started. It wasn't long after that Brian just sort of woke up. And it was everything anyone could have wanted. He was shocked and overjoyed that his mother and brother had come all this way to help him. He couldn't believe he'd ever take so many pills. Brian was delighted to be alive.

The three stepped off the plane at the Minneapolis–St. Paul International Airport at 10:32 p.m. on August 5, over a month since it all began and an hour and half before I entered my next decade. Jason, though clearly tired, looked strong and confident.

He'd become leaner during the ordeal, and while he had

dark circles under his eyes, there was something that sparkled in him. Something present. I'll admit, for an old married guy, he was pretty hot.

As we walked out to the minivan, holding hands, he turned to me. "Bird, Cambodia really showed me one thing. We're all going to die anyway, so why be afraid? And before I go, the only thing I want to do is—open that chicken farm."

As I looked up at him, my head worked against this new knowledge, the undeniable fact that Jason, this newly invigorated man standing in front of me, was never going to fit back in an office. It was clear he was completely and utterly non-cubicle-able.

And behind my smile, a seven-syllable version of the F-word bellowed in my head.

Chapter 3

One early morning in that planning winter of 2011, Jason's frustrated groan reached me back in the kitchen. It was the sad sound of a man trying to procure chickens.

I'd been refilling coffees as he read e-mail, seated at his antique secretary—a piece of furniture acquired specifically to keep Jason's unruly paperwork out of sight, but which no longer closed given all the poultry-related books and papers stacked on it.

He leaned back in his chair and absentmindedly pulled his hair with both hands.

"Hit another dead end for pullets," he said.

Pullets are the tiny first eggs of a laying hen, but it's also agricultural-speak for immature chickens not yet laying eggs. We were looking for just under two thousand of these young lady hens to be raised to around sixteen weeks, the cusp of sexual maturity known in the industry as "the point of lay."

"Can't we just raise them from babies?" I asked. Naively, I'd thought we'd be getting chicks, as we had for our home flock, picturing the children playing among heaps of chirping yellow fluff balls. There are mail-order companies that will dispatch peeping parcels directly to your post office, I told him.

"No," he said, taking his mug from me and setting it on a spate of papers titled *State Rules for Egg Processing and Sales.* "We're not set up to brood birds." He explained the huge equipment investment, starting with a good-sized barn, all insulated and rigged with heaters and fans and sensitive alarms that go off when the temperature varies by even a degree.

"Plus someone has to be up checking on them, around the clock. The truth is we need birds close enough to laying so that we'll have cash flow as soon as possible."

And we needed this cash flow, because by late summer Jason would be giving up his job. He'd been working at our local nature center writing its grants. It wasn't quite full-time and not quite forever, either. It was grant-funded itself, and layoff was imminent. As far as Jason was concerned, this was the perfect scenario as the farm ramped up.

I was still trying to convince myself that any of this was perfect.

Now, it's easy for someone from Big Ag or, in our case, Big Egg, to purchase, say, thirty thousand point-of-lay chickens. Just like it's no bother to go to the feed store and snatch five chicks from under a heat lamp for an urban flock. But what's not so easy is to find a brooding house willing to work with bird numbers between the tens of thousands for industrial-scale egg factories and the literal tens wanted by small-scale hobby farmers. We knew because they'd all turned Jason down.

Nor could we call on the broad network of midsized farms that would have the know-how and space to raise a couple thousand hens for us. Back in 2011, this simply didn't exist.

We were stuck in the middle.

This theme of being too big or too small for any of the conventional paths would haunt us over the years to come. We were still several years out from discovering the middle niche we occupied and its economic and social importance in the agricultural spectrum. Back then, we weren't champions of rebuilding rural America. We were simply screwed.

Given that Jason's plan required 1,800 nearly mature pullets by midsummer, the only option was to find a freelance operation willing to contract with us. After exhausting all the contacts of his contacts, Jason took the great pullet search to the Web. And while there were a few places in Pennsylvania with mature hens to sell in smaller numbers, here in the Midwest we came up with exactly one guy—Myron, who sold farm equipment in Iowa and raised birds on the side. He agreed to brood our hens and we, the insecure chief officers of the newly named Locally Laid Egg Company, breathed a sigh of relief.

But that serenity didn't last long.

After Myron agreed to our order and cashed our check, he wouldn't return our calls. Jason would leave messages, texts, e-mails looking to get some basic questions answered, or an update, and . . . nothing. Among the many business phantoms that would wake us in the night, Myron began playing a starring role in our nervous pillow talk. We may not have known much about starting a farm, but we understood it would be tough to launch an egg operation with no hens. I mean, it's not like we required a photo album with baby pictures of these chickens (okay, we actually would have loved that), but we had some questions, and a simple reply, any reply, would have gone a long way. This worry drove Jason all the way to Iowa—literally.

As he tells it, he pulled into the driveway to a chorus of barking dogs in kennels and the rural ornamentation of broken machines and vehicles. With no one around, Jason walked the property, taking in some of the open barns filled with poultry incubators and bird netting.

"Eventually, Myron came out of his house," Jason said, describing him as a man caught off guard despite all of the efforts to announce the arrival.

There was a defensive flurry of excuses why all the messages were lost. Myron even blamed his assistant for not alerting him to Jason's missives. He took Jason inside his office and pulled out our file, where Jason's letter sat on top.

Fortunately, Jason is a talker. He can befriend a store mannequin and makes me, an ebullient lover of people, look like an introvert. His animation stems from a sincere interest in everyone's story and a love of asking questions. Often, too many questions. For this excited quality, I call him the Golden Retriever of Husbands, and sometimes I must gently remind him that our new acquaintance is not on trial and that he should let him or her finish answering his last query before asking another. Fortunately, Jason's good nature shines through and like a bounding happy dog, one can't stay annoyed with him for long.

Myron and Jason then crossed the yard, passing large

warehouses until they reached a smaller building, a brooding space. There, Jason met our hens.

As idealistic as he'd been with our backyard flocks, a collection of chickens that perk up at his voice, land on his shoulder, and, I'll say it, flirt with him, it's hard to overstate this moment.

This was the first time Jason had met unsocialized birds. Chickens who'd never been handled, much less enjoyed, Beatles songs sung to them. Having had minimal people contact in the warehouse, these birds had bonded only with each other and functioned in a sort of neglected collective. It's a safe guess that there weren't a lot of independent thinkers in that building. For these pullets, resistance was futile. If one chicken moved, they all moved. And when you're talking numbers in the thousands there's a natural delay of movement that mimics a swell in the sea.

We have friends whose daughters went to elementary school with our children. Katherine and Rachel, bright girls, have decided that each of their five backyard hens has only a fifth of a brain, evidenced by the behavior of those who lose track of their group. These chickens become agitated, flap their wings, and squawk ungracefully to get back to their think tank of sisters.

Following this poultry math, the chickens Jason met that day had $1/1,800$ of a brain, and it showed. When he walked in, the birds backed away on a wave of undulating terror swelling around the dimly lit room.

Beyond the cacophony, what Jason noticed was that these chicks had no food and water. It made his throat bitter to see this poor stewardship of our paid-for chickens. Not caring if it was bad form, Jason stepped in and started pulling the empty water-ers and feeders to refill. After a minute of watching, Jason said, "Myron got the message and joined in."

Later, while driving home, Jason phoned. "If we don't brood our own flock next time, I think I'll find someone else." I remem-ber resisting those words, not wanting to start at square one of finding a farmer willing to deal with us.

"Really? I know how hard it was to find Myron."

"Yeah, I think we'll go with someone else next time around."

This was said in a flat manner that told me Jason was wres-tling with thoughts, ugly ones, and protecting me from bad news. I opened my mouth to press further but at the last moment decided to wrap myself in the blissful Snuggie of ignorance he was offering me.

It seemed there was always something about the farm startup that required comforting in those days. I was still coming to terms with the unhappy fact that our flock wasn't going to be served organic corn—rather settling for a locally grown, non-GMO variety. This standard of sourcing and selling locally would prove to be more far-reaching than I ever could have imagined, but at the moment I was stuck on what we weren't. And that was certified organic.

When Jason had gone out looking for like-minded farms to help plan Locally Laid's model, he learned about two small egg operations with organic pastured poultry.

"Great!" I'd said. "When can you go see them?"

That was when he told me the disquieting news that they

were both out of the business. Word on the street is that fluctu-ating cost and availability of organic feed was to blame.

My first thought was: *Wait, we live in the Midwest.* I mean, this is the earth's grain epicenter. Why would there be short-ages? Apparently, like everything else associated with doing the right thing, it's more complex than that.

It starts with the rare crop farmer who's gone through the three-year rigor to become certified. It's a commitment that requires writing a series of checks, ones to regulators who travel to inspect the operation and others for the registering organiza-tion, all the while wading through a paper-heavy and often costly process as the land transitions into an organic farm. It's strict and I'd argue that it should be; it just means there are fewer folks willing to take that path. In fact, only about 0.7 per-cent of the roughly two million farms in the United States are certified organic.

Following a certified organic harvest, these grains must go to a certified mill that won't commingle organic with conven-tional crops. In our northern-tier location, that would mean dis-patching a truck nearly nine hours round trip to the farm to fill feed bins twice a week—releasing a Riverdance of carbon foot-prints on our low-food-miles ethos.

And that is, when it's available.

What would a farm like ours do if we couldn't source organic one week? This explains why there are regular shortages of organic eggs in the supermarket. The idea of ensuring that what's laid during a week of conventional feed ended up in car-tons with accurate labeling and bar codes would be a stomach-churning, logistical whirl. But putting aside the food miles and the factual labeling, the biggest sticking point is this: most of the organic grains used to feed animals in this country are—brace yourselves—from China. India, too. Despite the breadth and depth of America's breadbasket, we import as much as eight times as many organic grains as we grow.

Yeah, it's a sad truth.

Freighters carrying our conventionally grown kernels make their way to Asia as ships of Chinese and Indian organic grains fulfill our demand for more organically grown foods. In addition to the scads of food miles—some six thousand for Chinese crops—these imports have been viewed with suspicion and, at times, flagged for not meeting the organic standard under which they've been labeled. One has to wonder what *organic* really means when a crop is grown in some of the world's most polluted nations. Is shipping sustainable feed halfway across the planet to make consumers feel good about the environment truly environmental?

Even author Michael Pollan has said, "Local supports a great many more values than organic—even when the local food is not certified organic." He goes on to list them: the environment, less reliance on cheap fossil fuel, stronger communities and keeping nearby farms viable, saying that if these are your concerns, ". . . the choice is simple: buy local."

That helps.

I also take solace in the fact that farmers in our region will be planting more lower-yield, non-GMO varieties of corn because we said we'll buy it—and the money we spend on it will stay in our little part of the world. But even so, with all these solid reasons, I'm still dejected by the feed situation to this day. Just because I understand the "why" behind our feed choices doesn't mean I don't pine for better.

"And remember, Lu, that certification just refers to the feed. It's not like organic eggs are from birds that go outside," Jason said. "It simply means they get to eat organic feed inside their warehouse."

I churned this all around in my head until I reminded myself that two chickenless farmer wannabes couldn't solve it all. In fact, it likely requires the momentum of government subsidies. And before I get lectured on the harm of big government, let me remind you that we as a taxpaying people manage to fund food additives like high-fructose corn syrup, corn starch, and soy oils at the rate of $17 billion a year (that's a number with twelve zeros) while 0.01 percent of that number, $261 million, is used to publically subsidize apples. Imagine how many organic feed mills we could build, how many organic certifications we could grant with just a small percentage of what we spend on, well—crap?

Why we don't, I speculate, has to do with the toxic soup of big corporations, campaign donations, and lobbyists wielding political influence to keep those agricultural dollars directed exactly where they've been going for decades.

All this made me sigh. It seemed that starting the business was a series of disappointing compromises and heart-stopping leaps of faith.

Chapter 4

Over a nine-month period, Jason had taken business and farming classes and even visited a real pasture-raised egg producer in Maryland. Of course, I'd hoped he'd see the enormity of this poultry endeavor and decide, all on his own, that it wasn't feasible. Instead it meant some additional knowledge for Jason and a little more time for me to acclimate.

But not enough.

He surfed Internet sites every night looking for acreage, which, following census trends, wasn't easy to come by. Farmland values have steadily increased over recent years, some 748 percent since 1987 to 2014. U.S. land prices have bloated as countries around the world have risen in affluence; when a population starts to have more money to spend on food, they want meat. Meat takes corn; corn takes land. This cycle effectively shuts out cash-strapped new farmers from owning property.

We couldn't afford to buy, so renting it was. And despite Jason's trolling of Craigslist and real estate sites, he found land the way people have for decades. He went to the feed store and started a conversation.

That late winter, with gray skies pressing the wet snow to the ground, we drove half an hour south and turned onto a country dirt road. Another left later, we crunched into a rock driveway, the nose of our minivan positioned directly in front of an aging pole barn. Looking at it, I guessed that it was once an open-air structure that was later closed in with red-painted particleboard. Now it was regressing to its original state with the odd board missing—like a puzzle shedding pieces.

Jason gazed out the windshield with an excited grin and narrated the Technicolor movie clicking on his internal projector screen. And from his running commentary, it was as though we were seeing two different properties.

"We are so lucky there's a barn. We'd really only be renting the land, but the landlady, she's cool with us storing our construction materials in there while we build. Plus, there's a tractor she says her boyfriend could help us use sometimes."

I was underwhelmed, a quiet foil to Jason's animated chatter.

"And you see how open the pasture is?"

He waved his deerskin glove to the right, indicating an expanse of open field, some 150 acres below a layer of unenthusiastic snow. "Just think of all the chickens we'll have foraging out there." He said this with a tangible longing.

Though I wiped the condensation away from the van window with my mitten and squinted, I couldn't see them. I don't think I wanted to see these birds whose existence would mean the end of any typical life I'd once envisioned for my family.

A wise philosopher, or it may have been an Internet meme, once said, "Choose your choices." But try as I might, I still couldn't quite embrace this farm that we were choosing a little more every day. Instead, I lived with it the way I imagine one lives with a grizzly bear in a campsite. Never looking straight at it, keeping it in that part of your vision just outside your center of attention and hoping it just might harmlessly lumber off in the underbrush. Or you know, get hit with a high-powered tranq gun and teeter off a cliff.

So a quiet part of me, one I revealed to few, still wished this whole idea of opening the first commercial-scale, pasture-raised egg operation in the upper Midwest might just drop off. That's why I couldn't tune in to the all-chicken channel Jason was watching on the snow-covered pasture.

Or maybe I couldn't picture frolicking poultry because I couldn't square up the unvarnished realities of a working farm with the bucolic barn photos in the free calendar on our refrigerator. Taking in this operation, a place with its fair share of deferred maintenance, the only thing I could make out through the dreary gloom giving into evening shadows was hardship.

Growing up in Maine, I remember driving past the peeling signs of an abandoned farm collective on a back road to the coast. It was part of the larger back-to-the-land experiment of the 1970s, when hundreds of idealists attempted to emulate Helen and Scott Nearing's homestead, profiled in their bestselling books *Living the Good Life* and *Continuing the Good Life*. These were manuals really, outlining their self-sustaining and intellectual lifestyle along a picturesque bay in Maine. Many who followed failed miserably in their attempt to grow food in a climate where one wakes up to frost two-thirds of the year.

The descriptor *miserably* felt palpable to me.

I thought of the young and not-so-young couples who cleared land, built basic shelter, hand-dug a well, or put up cold frames to start seedlings in the early spring. Not surprisingly, divorce and poverty all but squashed this homestead campaign within ten years. This, too, was palpable.

However, the Nearings weren't the only ones to write about the movement. Others have published books pointing out the famous couple's advantages along their "Good Life" road. This included supplementing their cash crop of blueberries (and before that, Vermont maple syrup) with sizable inheritances, an abundance of volunteer labor from eager farm groupies, and a lucrative spot on the lecture circuit. The latter not only paid them but ferried the couple away from their stone farmhouse during Maine's harshest months.

While we couldn't count on trust funds or speaking fees, we did share one element with the legendary couple—an uncommon generosity of friends. It's hard to tell if people were drawn to our project's idealistic sustainability and largish scale or perhaps just the pluckiness of the whole darn thing.

Locally Laid's elite "kitchen cabinet" convened around the dining room table of Gail Blum and John Erickson, architect friends with some farming background who took us on as a project, and included Rod Graf, a skilled carpenter and designer. Looking at the group, one could see these are people who know how to use their hands and smarts to solve problems, fix things, build things—to tackle situations Jason and I have generally bought our way out of. I cannot overemphasize their importance to the farm venture.

They were there to help Jason answer the manifold questions a project like this entails, many he did not even know to ask himself. They also wanted to ensure that Jason really knew what this big commitment was—the business equivalent of poultry matrimony.

Later Gail would tell me that Jason had to be educated. Influenced by books he'd read warning against spending too much money on any farm need, he'd suggest constructing things out of fiberglass cloth and duct tape, the materials from his sailing youth, but nothing that would stand up to the daily wear of agriculture.

The group resisted collective eye rolls and calmly explained the laws of nature and physics that prohibited Jason from doing things he wanted to do. I can only guess their generosity was out of quiet worry that without their counsel they'd be forced to watch us lose everything.

It was a valid concern.

They made a plan for poultry houses that allowed for easy pasture access and the building of a processing facility to wash eggs to state standards. Jason would return from these meetings, often lasting upwards of three hours, adrift in details and under pressure to cobble together answers bereft of experience.

Gail even built a fantastic cardboard architectural model of a wooden coop, complete with nesting boxes cleverly accessible from little doors on the outside. It was one of the prototypes the group auditioned, and later even built, but the design was nixed for lengthy construction time and high cost.

To me, not nearly enough time had elapsed between my marveling over the charming model coop in the comfort of my Duluth living room in those early days of 2012 and starting to build real ones on that rented land in April.

I stood that cold Saturday morning shivering in the barn waving to our kind friends who had carpooled the thirty-five miles from Duluth to Wrenshall, parking on the gravel driveway. As they walked past the white two-story farmhouse where our landlady lived, I could see that their attention was drawn to the other side of the drive, where acres of field gently rolled out to a far-off tree line. We'd rented 10 of the 150 acres, some already let for beef cattle and hay production, allowing us scenic cow and grass views.

All the fields lay just beyond a set of two metal cattle gates. This setup created a fenced-in buffer zone between the pastureland and the driveway to prevent one from accidentally releasing a herd of cows into the rural landscape.

During the weeks since Jason had shown me the site, the pasture had transformed. The receded snow revealed hesitant chartreuse shoots peeping out from the overwintered grass, accompanied by an earthy pungency that one can almost taste in the air.

And he was out in the middle of it all. A good thousand feet beyond the metal gates, Jason was working with his own volunteer crew on open pasture, constructing the first of six hoop coops to serve as home to our brood. These rectangular structures, at 120 square feet, were smaller, portable versions of the temporary semicircular tarped houses that spring up overnight in grocery store parking lots to hawk annuals and perennials. It kind of looks like a tunnel made of tarp with chicken wire on the ends, tall enough to easily accommodate a people-sized screen door.

Despite the cold drizzle, which later turned into pelting rain, we watched a group of four men and two women wrestle one coop's metal U-shaped frames into an upright position. These would be fastened together with others and later skinned with an industrial-weight white tarp to form the oblong shelters with room for some three hundred hens apiece.

Soon, the elements drove us into the barn. The early-spring weather was seasonal, which is to say damn cold with a 100 percent chance of craptastic. Off-and-on freezing drizzle punctuated with the occasional driving hailstorm: this is springtime in the Northland. Where we sit on the north forty-sixth parallel, just a few klicks down from Ontario, we're slammed by both Canadian cold fronts and the capricious whims of nearby Lake Superior. Each on its own is formidable. Together, they're a ferocious duo, turning a season known in most parts of the country for sporting light Easter dresses and frolicking among daffodils into cold, wet misery.

The building party probably should have been canceled that day, but there was a schedule to keep. Our first flock of nine hundred chickens was set to arrive in June, and the second, same-sized bird delivery would come approximately a month later.

So Gail, John, and Rod, along with a dozen other folks, gave up the comfort of nestling in at home to haul their power tools and hefty generators in the beds of their capable trucks. They spent the day framing out hoop-house kits we'd purchased, each seeming to come with a mysterious flaw only discernible during the build-out, often necessitating an unplanned run to the nearest hardware store some ten miles away.

Fortunately, the task that lay ahead for my group was seemingly more straightforward: building nesting boxes. We would attach thirty of these open metal boxes in each of the six hoop houses. Stacked two on top of each other in rows, these one-foot-by-one-foot open boxes would line the left side of the coop's interior and later be filled with some straw. Chickens would then seek out these quiet, shaded places for the privacy preferred when laying their nearly daily egg. I say "nearly," as a hen in lay will produce an egg every twenty-six hours, and at about day twelve, she'll take a little production vacation. An eggless holiday. (A poultry professor I know claims that this schedule puts chickens into PMS for one and a half minutes every day, but I've seen no evidence of this.)

In the barn, where we could still see our breath in small puffs, we started our task by scrounging for battered sawhorses for makeshift workstations. Then we quietly set about puzzling through the cryptic instructions provided for the metal kits we'd purchased from a company in Kansas. It was quiet for a while as we examined the illustrations and tiny print—that is, until the person who'd assembled the most Ikea furniture naturally took charge. The expert was my neighbor Karie, who wiggled and wedged the metal sheets in unexpected ways until they formed a container a little bigger than a breadbox (sans door), plenty roomy for a chicken or two.

It was then up to the rest of us to wrangle bolts through slightly misaligned holes to fasten the whole kit together, finishing with an ill-fitting nut. Our fingers protested the precise movements in the cold with an ache that oscillated between a dull numbness and prickly burn. The repetitive movements may not have produced heat but did make for some colorful swearing.

"Gah! This damn thing is nearly frozen in place," I protested while trying to straighten a would-be box.

"Yeah, but at least you won't have to lie naked in it," said Julia Singer, an accomplished poet and great observer of truth.

Production stopped as we all turned to look to her.

"Well, think of the hens," she stated baldly.

In near unison, we turned our heads back to stare at the metal crates in front of us. Pulsing cold would transfer from the hard steel during our long winters, just as surely as the summer's heat would singe feathers. I concluded that I wasn't going to be the only female with complaints about this venture.

Despite these boxes being made of metal, it truly was a throwback setup with more in common with century-old egg production than modern operations. According to the *Poultry Tribune*, there were well over five million farms with egg-producing hens in 1900. By today's standards, these flocks were small, minuscule— maybe a couple of hundred hens roaming the farmyard eating waste grain, weeds, insects, and food scraps. The egg income from these flocks, tended almost exclusively by farm women, was called "pin money," so named for being attached to the inside of a dress bodice with a straight pin. I like to think of it as the precursor to the modern "mad money"—cash hastily jostled into a bra before heading out on a date.

Of course, this was just cottage industry stuff. Birds were in a stewpot come late fall, as winter egg production wouldn't even pay the cost of feed. It wasn't until 1928 that research published in the *Northwest Monthly*, a publication issued by the Northwest School of Agriculture at the University of Minnesota, changed everything. Discoveries out of the small town of Crookston centered on the use of selective breeding and cod liver oil—an extract chock-full of vitamin D—mixed into the mash and supplemented with electric light. Hens would lay right through the dark months—a seismic shift that took eggs from seasonal luxury item to everyday staple.

Chickens need vitamin D to absorb calcium, a mineral that can make or break an egg-laying hen. They stash their supply in their hollow medullary-style bones—think of it as kind of a bone within a bone that serves as a mineral store. When a bird starts laying, her calcium need jumps fourfold and without enough, she'll play the martyred mother, pulling it from her personal supply.

With this good news out of Minnesota, eggs now offered year-round profitability, and more attention was paid to the industry. Another breakthrough came ten years later when Milton H. Arndt, Illinois farm boy turned New Jersey inventor, created a stackable birdcage concept akin to a "filing cabinet in a modern office building." It worked where others hadn't, given that with vitamin D, birds, strictly speaking, no longer needed sunshine.

Arndt's nearly century-old setup greatly resembles modern battery cage systems. Present-day cages often house five to ten hens (the industry tends to use Leghorn chickens, a smaller bird) in a wire crate often 2.25 feet by 2.25 feet and 14 inches tall. That works out to about 67 square inches per hen. For comparison, a standard piece of printer paper is 93.5 square inches. This isn't enough room for a bird to stretch out; even a petite leghorn has a wingspan of 26 inches. Nor can they indulge in other instinctual behaviors like dust-bathing (essentially digging a hole and flopping into the cool dirt to soothe the skin under their feathers) or, you know, walking.

But what these models did allow for was more eggs with fewer laborers, a siren song to industrialists everywhere. While most businesses had long succumbed to reductionist systems of greater mechanization (think Henry Ford's assembly line), agriculture had been the square peg. The devil wasn't so much in the details as in the variables. But now that a building could control for weather and predators, and even regulate how much feed each bird could access, it was ripe for expansion.

Battery cage systems caught on nationally in the 1950s, and consolidation in the industry began. If we scroll ahead to 2015, the American Egg Board lists a mere 268 companies accounting for 95 percent of the nation's 305 million laying birds. Seventeen of these companies have flocks of 5 million chickens, though *flock* doesn't seem to be the right word for numbers that large.

The number of egg producers has been reduced a stunning 99.99 percent in 110 years.

Heck, it's an 89 percent decrease from the 2,500 egg producers around in 1987.

That cold day in the barn, I remember being grateful as lunchtime neared and I could stop fretting about frigid hens and step into my true role at these coop-raising events: feeding people. It's a task I gratefully shared with Mimi, as Jason's mother is the better cook between us.

I readied the makeshift barn buffet, a plank supported by two cinder blocks, for our slow cookers filled with pulled pork for sandwiches and baked beans. We weren't big meat eaters ourselves, maybe serving it once or twice a week. This stems from me, the primary cook, having read the famed *Diet for a Small Planet* back in the 1980s. This small missive laid out the resource intensity of meat production with its high water and grain needs. So while I found meat tasty and, honestly, more socially acceptable to eat at other people's homes, I generally served a simpler, more plant-based diet than the typical American household.

But today, our wet, tired friends deserved the hot, rich pork. I was glad for Mimi's help. She carried in a plate of her homemade chocolate walnut bars as Abbie and Milo shadowed her. Not just for her treats, though that helped, but because she provided them the kind of full-eye-contact attention that was a rarity in our busy farm-building household those days.

I was also glad she'd driven the two-plus hours up from her Twin Cities suburb because I knew she'd make sure this meal wasn't just hearty and good, but also lovely. She'd found pretty paper plates with matching napkins, remembering little details like dessert forks and cream for the coffee. Despite being served over a dirt floor, this farm-elegant meal conveyed our sincere gratitude to these cold volunteers literally building this dream. It was difficult to accept all this freely given generosity.

Putting out the last of the rusty folding chairs that propagated in barn corners, I couldn't help but think the luncheon had the air of a shower, an event commemorating a big life change. Sitting down, we formed a loose circle, plates on our laps, while our supportive friends, many of them business owners themselves, murmured encouraging words to us.

To be truthful, I've grown suspicious of life events that trigger showers. It feels like the calm before the storm, the harbinger of things to suck. Historically, these were occasions for women to share their collective marriage or child-rearing wisdom gathered along their own journeys. But that's not what happens today. We've become too politically correct to issue opinions based on our experience, thus leaving attendees of such fetes to fall flat of the original intent. I know; I've participated in such group failings myself.

But unable to bring ourselves to lay out reality for the honoree, we adopt an "ignorance is bliss" attitude and distract the guest of honor with a Cuisinart, a Diaper Genie, and assorted petit fours—and, like those gathered around the barn, just smile, hoping for the best for this new endeavor.

Chapter 5

2012

It was summer in the Midwest, which is to say it was idyllic.

Long days rolled out like the land before us, expansive and dotted with wildflowers and fireflies. It's a potent season that can enchant away all thoughts of our protracted northern winters, which seem not only far off but altogether improbable. I blamed it on the heady billow of chlorophyll and vitamin D. In this unfiltered sunshine even a non-agrarian like me could see that this swelling farmland was beautiful and precious, a ripe expanse of great potential.

I just wasn't convinced it was our potential.

But we were about to find out. With a transition as subtle as an ax, today would come to have a story, a moniker, and forever more be *The Day the Chickens Came*.

Turning my head, I picked out a cow in the adjacent field making direct eye contact with me, chewing cud and staring through my sunny-day facade. Perhaps she caught the acidic smell of fear wafting over the fence. Or maybe I was just such a spectacle of floppy arms and loose ends that even a bovine couldn't help but stare.

In spite of our many weeks of preparation and today's clear skies, I felt like a pratfall in the making. My head spun with thoughts like, maybe there's a reason why we're the FIRST commercial-scale pasture-raised egg farm in the ENTIRE upper Midwest.

I was a farm fraud, a prairie train wreck.

Pocketing a box of industrial staples, I scrambled over the second of two cattle gates rather than fool with their rusty locking mechanism. Though it was early, the metal was already hot, but on the other side of the fence, the pasture smelled like happy childhood, a fragrant mix of sweet grass, warm earth, and clover stirred up under a kid's sneaker. I breathed deeply, trying to steady my nerves.

I looked toward the hoop coops, just under two football fields away, set on the pasture's highest spot. Seeing the six oblong structures, white-tarped and complete with roosts and nesting boxes, even I had to admit they looked beautiful, even chicken-ready. Around each one was unspoiled prairie, cordoned off with portable, solar-electric fences necessary to rotate the birds onto fresh grass.

The best way to understand the revolving pasture process is to mentally draw a circle representing a couple of hoop coops, housing some six hundred chickens, and imagine it as the center of a flower—like a daisy a child might sketch in crayon. Then draw a simple U-shaped petal off the coops. This would be the first pasture run for the birds, also called a paddock, created with the flexible fencing. After a couple or three days, as the hens roamed within this area freely, picking down the grasses, thinning the herds of insects and foraging every single seed, it would be time to erase the first petal and draw another adjacent to it. This would be the fence line for the next fresh space for the chickens to roam. And so it goes, until the birds have migrated all the way around the coop. Some two weeks or so later, they return to that very first field, which would then be verdant and ready again for their fervent poultry assault.

That is, when the birds arrive.

"Lucie!"

Jason was waving me over to the first chicken house. There were well over a dozen people on the field now. Some were folks Jason had met in his agriculture class; others were friends whose children went to school with ours. While they didn't all know each other, they shared a common excitement to be part of this big, idealistic thing from which they could walk away freely. I don't mean that to sound ungrateful; it was actually envy talking. Part of me longed to sneak into their tidy cars and fold myself into their sensible lives.

Jason introduced me around, and while I smiled and shook hands, I wasn't registering names.

We did little tasks as the hours passed in the hot sun. Answering question after question regarding our farm plans made my happy mask of "Isn't this great?!" pinch my face. I needed a minute alone. The opportunity arrived while bending wire back around a door frame. I'd taken off my leather gloves for a better grip on the stubborn material when a sharp end found a tender spot under my fingernail. The quick jab bled persistently, requiring a Band-Aid.

I pawed through the first-aid kit in the back of the Kubota, a glorified ATV built for farm use. It was the first of many purchases not originally built into our business plan. These included several generations of waterers and feeders and, oh yes, a several-thousand-dollar well, but that would come later. The ATV's four-wheel drive and hefty suspension became necessary as spring transformed the field into a sucking clay trap that made transporting lumber needed for the floor frames of the coops with the minivan (our most rugged farm vehicle) impossible.

The gift of good credit allowed Jason to walk into a dealership and, after penning his illegible signature on a $10,000 note, walk out with next-day delivery. The four-wheeler was happy orange and easy enough for our children to drive with supervision. They called it BoBo. In my head, I referred to it as the "summer tour of Europe I will never have," but I tried not to dwell on it.

There was little aid to be had within the filthy Red Cross kit I found in the back of the little vehicle. More Band-Aids were in the minivan, I remembered, allowing me to depart the chatty group busy filling the last of the pristine waterers, a sort of inverted plastic jug that drip-flowed into a red lid for easy hen access. Holding pressure on my cut finger, I headed for the big field gate and awkwardly waved at our children, Milo and

Abbie, then nine and eleven, who were looking for bugs in the grass to give the chickens when they arrived, something they've done dozens of times with our backyard birds.

Just before the fence line, I veered the few steps over to the frog pond. The heat, now bordering on overwhelming, made even the sight of this stagnant water inviting. Reflexively, I checked in on the tadpoles, a daily highlight during these past few weeks of construction. I looked to see if any had started to sprout legs or dropped a tail, but I was careless with my shadow, giving myself away. They flitted to their hidey-holes under the marsh grasses.

I want a hidey-hole, I thought, just as Jason loudly snapped his cell phone shut. "They're almost here," he shouted, announcement style, and a murmur of excitement passed over the farm. I hustled to tend to my cut and repositioned myself in the field before the big arrival.

We'd decided that our chicken supplier, Myron, was nearly there. That morning he'd spent hours crating up the birds for travel and was on the road before checking to be sure that today would indeed be a good day for us to take on a trailer's worth of wriggling, pecking, scratching, clucking life. Had he phoned, even we nascent chicken farmers would have told him not to transport birds today.

It was the weekend of Grandma's Marathon, an event that annually draws some ten thousand athletes and twice that many spectators to Duluth. The commotion clots up the region's highways and side roads. Add to that the fact that we were experiencing a rare heat wave, and that meant a truck full of hens, *our hens,* would be sweltering in stop-and-go traffic up the normally free-flowing Interstate 35.

I heard it before I could see it. A truck approached the farm at high speed, kicking up dirt on the gravel road. It was the king-cab type, the kind that carried two extra wheels under full-fender hips. And I was surprised to see it hauling a flat, open trailer with scores of red and yellow crates fastened to it with industrial strapping and bungee cords. It was completely exposed. This was not the ventilated semi we were expecting, but rather a two-bit operation held together with spittle and bailing wire.

"Oh," I whispered, my hands pushing my hair on top of my head. "He's a bigger yahoo than we are."

In my chest, an icy spot called all the blood in from my extremities. I was lightheaded. I was mouth breathing. I'd read this is the body's response to stress, readying the large muscle groups to fight or flee. But my sympathetic nervous system was actively engaged in the least glamorous third option—freeze. I was a dull human paperweight on the prairie.

As everyone sensed that something was amiss, the energy chatter drained away from the field. We were all staring. When the rig stopped, Myron—a large, sturdy man—leapt out, yelling and waving his hands. Because I was a good five hundred feet away, I couldn't make out the words, but his tone and big gestures require no translation.

It was panic.

He gestured wildly toward the metal gates that led onto the field. We stood, stunned and unblinking.

Clearly, there was trouble. Myron shouted through the repressive heat, yelling over the noise of his truck towing the open trailer hauling our nine hundred chickens. It set Jason into an open sprint from the farmhouse driveway. The words echoed through the stifling air. From where I stood in the field, I couldn't tell if they were yelling to or at each other, but either way Jason made quick time through the heat to the first set of big swing gates.

After months of working here, he was practiced at lifting the off-kilter metal fence with his foot to ease the rusty chain from its locking mechanism. And before Jason had run just over two hundred feet to the next one, Myron had gunned his big rig with its long, open trailer through the rutted and marred path between the barriers.

I held my breath and felt a small trickle of sweat down my neck. My fists opened and closed in a seemingly involuntary manner.

Jason made it to the second gate for his next foot-balancing, wiggle-chain dance and pushed it open just as the truck sped through. It appeared that Myron did not slow down at all. I swore under my breath and was surprised as Milo's hand stealthily slipped into mine.

I gave my son a little squeeze, looked down, and smiled. "Wow, it's an exciting day, isn't it?" His blank nine-year-old face told me he wasn't buying it, either.

The truck roared to where the chicken coops sat on the pasture, and Myron was out of the vehicle so quickly he left the driver's door open. Insistent beeps emitted from the cab. Yanking tow straps, he shouted orders to grab poultry crates. As I approached, the urgency became clear. These birds appeared limp in their travel containers, which looked like narrow and rectangular four-sided versions of milk crates with a latch.

Our assembled friends emerged from their stunned states and were now passing chicken containers fireman's-brigade style. Myron moved to the front of the line at the doors of the hoop coops and cut open the crates' zip ties. He reached in with his huge workman's hand, grabbed hens from any place he could get a grip, and flung them into the coops.

I was stupefied.

Our experience handling chickens involved the utmost respect: a gentle grasp around their body to avoid an upsetting frenzy of wings, and then a quick tuck under an arm where perhaps they could hear our hearts. We had always cooed and spoken softly to our birds, looked gently into their faces in search of an interspecies connection, and whether real or imagined, we'd always found one there.

In contrast, Myron hurled birds, making a pile of broken ones to kill later.

I willed my body to move to start de-crating hens, because every bird I handled was one less that Myron would touch. I didn't want him touching any of our chickens. Just as I leaned forward, I heard a terrible cry and turned. It was my Abbie, her skinny-kid, preteen body dominated by her open mouth, contorted with shock and anger as she took in Myron's tyrannical acts. Tears leaked down her face, red from anger and the extreme heat. She was too upset to form words, but nonetheless she was pleading with me. Everything about her screamed, *Make it stop.*

I embraced her to my chest, then pushed her out just far enough so she could see me when I spoke.

"The chickens are so hot that we must move very quickly to get them into their coops, where it is cooler and there is water," I said. "They will be better once they're inside."

Still, her eyes weren't on me; they were looking over my right shoulder to Myron, who was moving like a bulldozer through the crates.

I gave her shoulders a gentle shake. "Look, this will be okay." She surrendered her attention to me, but her expression said she would never again believe me.

I sent her to find her grandmother and picked up the next crate, circling as close as I could to Jason. We shared a look, a kind of a head tilt with raised eyebrows toward our daughter.

He gave me a curt, understanding nod. After two children, four moves, a military deployment, and a decade together, we've earned this form of nonverbal conveyance.

I watched as he walked up to Myron. Jason had been losing weight during the couple of months of construction, and Myron appeared bigger in height, heft, and persona. At that moment, I thought Jason brave, doing something I'm not sure I could— approach this rough and seasoned farmer with our bleeding-heart, newbie ideals of how to handle livestock.

Though I couldn't hear the conversation as I made my way to the trailer to grab the next crate, I could read the body language. Myron, though offended, stepped back—and acquiesced to Jason as the check-writing customer, not peer. The next chicken he picked up, he did not pitch.

Snapping open a borrowed pocketknife, I cut the plastic ties on the next crate. The birds inside had contorted to fit the shape of their low box, and I couldn't imagine what this long journey, exposed at highway speeds in the heat, had done to them. But rather than rejoicing at their freedom, the birds cowered in the corner and made me reach in past my elbow to grab them. Wrangling a gentle grasp on a chicken, I was alarmed by her size. She was supposed to be a sixteen-week-old hen, robust and nearly at peak maturity. Rather, I felt bones protruding under wispy feathering.

As I scooped her out, I saw that in addition to being stressed and skinny, she was wounded. These chickens did not have trimmed beaks, a common practice in large-scale poultry production. These nervous birds had been using their sharp beaks on each other.

Honestly, when we'd finally found someone willing to take our bird order, we didn't know about trimming, and if we had, I'm sure we wouldn't have requested it. There are varying accounts, depending on your source of information, about whether squaring off the beak-tip cartilage with a laser is painful. The only thing I can say with absolute certainty is it's controversial.

Months later, a farmer would sell us a few hundred older hens with trimmed beaks and while the fighting was less, these chickens were also less adept at grooming. I now know it's vital for a bird to be able to nibble mites off her feathers.

I also know that chickens are hardwired to peck—the ground, grasses, and, when anxious, each other. When they do, they expose blood, and where chickens see blood, they are compelled to peck more.

Partially, it's the pecking order at play. A flock is a social system with a hierarchy. There are top birds, which get the all-access-pass to fresh feed, the coziest roosts and nesting boxes, even the softest spot in the yard to kick up a dirt bath. This social system also keeps a flock physically strong. If chickens detect an illness in a sister bird, they will put her down. If they're stressed, their small poultry brain determines that having fewer chickens around might alleviate that pressure.

This problem of bird-on-bird violence is not a new one and, in fact, is responsible for the term *rose-colored glasses*. In the early 1900s, tiny eyeglasses were invented, even patented, for hens to wear to diminish peer damage. The December 1938 issue of *Popular Mechanics* had a sidebar about these red-tinted aluminum spectacles. The article claimed that they "stopped the fighting instantly" because they cut the chickens' ability to see blood.

Chickens, for all the bucolic imagery of pecking corn thrown by gingham-aproned farm women, are not vegetarians. Left to their own doing, they'll completely void a hapless frog into a change purse in a matter of minutes. The glasses sold by the millions, though they were later deemed ineffective.

However, standing as I was with a badly injured hen in my hand, I understood the temptation to try anything to prevent such violence, colored eyewear included. Almost all the birds in this crate had significant wounds, and all but one had had her tail feathers ripped out, leaving only bloody backsides.

The bird I held was alive but sported a gaping cut on her back so deep and wide I could have easily worked my thumbs in and, it occurred to me, break her apart like a deli chicken. It was hard to imagine a scenario in which she would live.

I scooped up others and brought them to their hoop coop, leaving the wounded one in the shade for Abbie to collect. She'd already set up an infirmary in a coop that wouldn't be used until the second batch of birds arrived next month. Gathering up the piles of bloodied birds, she brought them to this quiet space where she'd assembled water, feed, and handpicked clover. Our medical expertise was limited to cleaning wounds with water and a light mix of hydrogen peroxide, then slathering essential oils on their lacerations.

When all the chickens were in their prairie coops, we stopped, exhaled, and took in the scene. They moved about, pecked at bits of straw, and made some furtive contact with humans.

Though battered, they seemed all right. Now, to keep them that way.

ACT 2

Cluck!

INTERJECTION: expression of surprise, extreme displeasure

Chapter 6

Tension soaked into the grass as we took in the sight of hundreds of hens in their new coops and out of immediate peril. I even risked the thought that they would have better lives on our farm. This was a desperate hope of mine, that for all the work and risk we were taking on, we'd at least be freeing chickens to better lives. Ones with prairie access and room to completely open their wings into the wind. I momentarily thought that they might even be grateful for this stressful day that brought them here, but then again, that would be a pretty complex thought for a chicken.

As the afternoon was turning into evening, Jason took me aside and asked me if I'd go home tonight via Superior, Wisconsin, and pick up packets of a restorative water additive to bring back to the farm the next morning. Annette, a local hobby farmer, had recommended it for our stressed birds.

"I know Dan's Feed Bin has it," he said.

Checking the time on my cell phone, I saw I'd have to move to get there before it closed. I turned my head to find the children

but saw their slumped little bodies walking toward their grandmother's car. They'd head directly to our house with her, and
really that was best.

"Okay, chicken Gatorade, got it," I said, and headed for my
minivan.

With all the wounded and dehydrated birds, Jason had
decided to sleep at the farm, which meant he'd bunk down in the
wooden structure built to be our future egg-washing house. This
was where all the state-mandated processes regarding the readying of eggs for market would take place—the washing, the weighing, the packing. But that was a few weeks away. The structure
back then was essentially an unfinished shell that provided shelter and a pleasant woody, just-built smell. He also had a Porta-
Potty out there, so I guess, in theory, he was set up.

As I pulled on to the county road, I couldn't have been happier to leave on this errand and head back home. Our Duluth
house, while not a fancy one, held distinct amenity advantages
of indoor plumbing and soft places to sit.

I settled in for the ride. This thirty-five-minute commute
from Wrenshall was certainly one of our farm's disadvantages,
but this spot had been chosen as a balance of zoning laws to
price. We had also chosen to rent because, well, that was what we
could afford, and before we bought anything we had to be sure
this new business model would work—had wings, so to speak.
But another of the deciding factors was me. I did not want to
move to the country and even more so, I didn't want to move my
children to the country.

In just a few months, Abbie would enter middle school, that

formidable experience formerly known as junior high. I knew it would not be a good cultural fit to jam her into a rural locale, population 306, where the classes had literally grown up together. Of course, Milo was steaming up right behind her with his own established social circle to protect. I couldn't tear them away from the place they called home, again. And even Jason understood that when it came to the children, I would be unwavering.

It was still light on that drive. In June, there are fifteen hours of sun, but if you count the twilight hours of dawn and dusk, it pushes that number over seventeen. This was all cosmic makeup for the dark winter when that number dropped to a paltry eight and what sun there was often filtered through milky white skies. But all that seemed a long ways off that hot night.

Pulling into Dan's Feed Bin, I parked near the sign. It's an enormous 3-D acrylic black bull suspended a good fifteen feet above the parking lot with the austere message, *If we can't feed it, you don't need it.* Here was where my cell phone woke up from its rural doldrum. It beeped alerts for several messages; one was from our new Locally Laid Facebook page, and while I unbuckled my seat belt, I opened it.

The message was from a woman I didn't know, which was a thrill in those days when someone not related to us acknowledged our new venture. Pushing my filthy hair out of my face with a hand of equally dubious cleanliness, I scanned her message. She claimed to really want to "like" our page but needed to know what would ultimately become of our birds.

Not even a full day with chickens on the field and we'd already hit the spent hens issue.

Spent hen is the cringeworthy term for a bird who no longer lays regularly as she reaches the end of peak fertility, or, as we've joked, henopause. In backyard chicken flocks, birds are often kept for years as they slowly taper production and may even come in and out of laying, much as human women do with their own fertility. But in the commercial egg industry, where a whole chain of people and organizations count on steady production, when a hen starts to flag off somewhere from ten to thirteen months after beginning to lay, she's considered spent.

If chickens aren't slaughtered in a season, there's an industry practice called forced molting. Raised naturally, chickens will molt all on their own, losing and growing fresh feathers during the fall as days shorten and temperatures drop. Production stops (or greatly wanes) for a few weeks and when it returns, the chickens surge with a new peak of egg production, laying larger eggs.

This response must be prompted in confinement setups by altering lighting and feed. In both caged and cage-free operations, chickens are kept at controlled temperatures in windowless warehouses. And as if they were living in a casino, they have nothing to signal the passing of time. With no change in light, there's no seasonal trigger to start a bird molting. As our birds would not only see the out-of-doors but live in it, they would certainly molt, allowing them to lay longer—though we were unsure if these chickens would lay at the same production rate postmolt, or even how long they would keep laying. It seemed that determining when a hen is spent was more art than science.

Fortunately, at that point, our spent hen decisions were likely a couple of years out, though we had started exploring ideas of having the birds processed at a state facility and donating them to food shares when that time came. But when I'd approached these organizations, it seemed too much of a hassle on their end. For one thing, we no longer live in a society that knows what to do with an old bird. Most folks don't understand that the chickens they see at the supermarket riding the rotisserie carousel are usually just a few weeks old, fifty-six days to be exact.

I'll give you a second to process that.

Most chicken meat presented on a foam tray under plastic wrap is from a Cornish Cross hybrid. These birds have been bred to grow fast, like breakneck fast, and honestly, that's not too far from the truth given all the health problems these chickens have from their rapid weight gain. The moment these chicks emerge from the egg, they're eating, and that's pretty much all they do until they're slaughtered two months later. If the bird is not butchered, the weight of its own breast meat will eventually cause its heart to fail.

There are many reasons this breed is popular with chicken meat producers. For one, they're half the price of other chicks and ready for profit-taking in a couple of months. That means something, if you care at all what the birds eat. Non-GMO corn, soy protein, and quality vitamin mix gets expensive fast, and if you're feeding a chicken, one that was twice the price, for an extra six weeks, well, that's going to make for a costly drumstick. And because Cornish Crosses are so young when butchered, they're tender. And tender is what the American public has come to expect from their chicken meat.

A spent bird, in comparison, is ancient history. As a chicken ages, its meat gets tougher, but especially so in the case of a pasture-raised hen who has been sprinting around a field. She's turned fat to muscle, which does not make for an appealing menu item unless one is willing to cook it for hours, even days. Hence another term from our collective foggy past: *stew hens.*

Stew hens (once a main ingredient in Campbell's Chicken Noodle Soup) are made soft with long, slow cooking in liquid with the bonus payoff of a more flavorful meat. It does, however, require a hefty investment of time, which is tough to come by between rushing out of work at five o'clock and driving kids to the soccer field by six thirty.

After reading the Facebook message, I sighed a bit and typed a benign and truthful answer. I told her that we did not yet have concrete plans and were open to her suggestions. While this was true, we did have a few ideas. For one, there are always folks eager to have healthy backyard birds, on the cheap, even ones not in their laying prime. These older chickens will still give some eggs, just fewer, and this way one has avoided feeding a non-laying pullet for four months.

Also, we have connections to an Asian community in the Twin Cities, and they've been known to buy birds live off a farm. Not for much, mind you, but given that these hens would be pasture-raised and eating an expensive feed, I'd like for them to have one more stop on the circle-of-life ride, so to speak. If it prevents someone from eating one less industry hen fed antibiotics and other dead chickens ground into feed meal, that would be a small victory. But as none of these options were fully realized,

I didn't think it fair of me to get into it with my would-be Facebook liker.

I pocketed the phone, walked into the feed store, and asked if they carried sports drinks for chickens. I voiced this like it was an ordinary request, like my whole life hadn't just gone over the edge into the unknown.

When I got home, the company Facebook page popped up on my phone again. Apparently my spent hen nonanswer was not good enough, as the messenger unleashed a tirade on me.

> You and others like you who CHOOSE to raise these animals these days really do have a choice on what to do with them when they are "spent." Money triumphs, however. We need new computers and new patios, vacations, parties, private school for the kids and the list goes on, so we steel ourselves and close our emotions in order to kill the chickens. We tell ourselves it's survival and look to the past for justification. Sorry, but I'm not buying it. We don't live that way any longer, and to my mind it is no different than slavery. Kind masters you may be; but in the end you will sell for profit, pat yourselves on the back for your kind benevolence to the dumb creatures and buy more.

My heart squeezed and I felt the big muscles in my legs contract. After having spent the last year worrying over how to best provide great lives for birds, I was incensed. I wanted to immediately spew surly points regarding the irony of my spending my day actually *working* with chickens while she was home, likely clean and well fed, in front of a computer terminal possibly toggling between berating me and playing Farmville.

But I didn't.

The gift of being in one's forties is the occasional exercise of judgment. Instead of firing a tirade back, I turned off my phone and tended to our home flock, gathering eggs, refilling feeders and waterers. Then I ran a near-scalding bath, drinking a glass of wine. As I eased into the tub, I crafted impossibly rude responses in my head. How could someone with no skin in the game be so opinionated about our farm? We were so exposed in this venture; I was convinced that Jason and I didn't have one ungamed follicle between us. I went to bed angry.

Getting to an okay place about comments such as these took years. It was (and sometimes still is) difficult not to take personally the angry agrarian demands of people who have never, not once in their lives, seen a real live chicken.

Early the next morning, in the predawn hours that others refer to as night, I fussed over a reply. I was as careful of what I didn't tell her as what I did. I withheld the information that chickens regularly live more than ten years and that if every egg farmer kept their spent hens in perpetuity, as she'd implied, there would be a land and corn crisis in this country. Nor did I educate her on sustainable agriculture, which isn't about natural end-of-life care for geriatric birds, but a quest to feed neighbors real food while treating the livestock and the planet better in the process.

Instead, I painted a picture of our chicken infirmary, Jason

sleeping on a borrowed army cot and his careful stirring of elec-
trolytes into their poultry waterers.

Later, when I read her ebullient reply, I was momentarily
happy until the end of her message where she admitted this: As
a vegan, she does not eat eggs.

Chapter 7

At daybreak that beautiful June morning, the day after The Day the Chickens Came, I called Jason. It was clear from the heaviness of his hello that things were not going well. Chickens in the background made low *ca-cawing* sounds, bubbling up over our conversation.

"The birds are so stressed, Lu." The chickens had been running up and down the length of the coop and piling in the corners—two, three on top of each other. Jason's voice caught. "I've been taking trash bags of dead hens out all morning."

The words clenched my stomach. I swallowed a small gasp.

While we didn't truly know these birds yet, we'd been excited for these sentient creatures to experience better lives, enjoy the hunt, scratch, peck, wind-in-the-wings full chicken-ness in our green field. We'd brought them here for better lives than the industry would give them. Did they really suffer all those weeks in Myron's warehouse and that hot, blustery trip on an exposed truck only to die hours before we would open the coop door to pasture liberation? It was enough to make us throw our heads back and let out an anguished, tonsil-exposing Charlie Brown "Aaugh!"

Then I took another breath and allowed myself to think about the money involved. Real money.

From our business plan I knew that each hen, if we kept her some twenty months, laying an average of 86 percent of the time, would produce some 523 eggs over her nearly two years. Of course, I now understand how off those numbers were. It was these idealistic projections that had Jason believing that during our first month of sales, we'd nearly break even and that by month two, we'd actually be making money.

In fact, it wasn't long afterward that our business plan was so off the rails that it reminded me of another fussed-over document that assumed the best outcome at every turn: my birth plan. Nowhere on that two-page, single-spaced memo was it written that I'd endure thirty-six hours of protracted labor, slap a nurse, demand drugs (street drugs, if necessary), and swear at a team of residents for "treating my stirruped legs like a scenic turnout." This is all to say, things do not always go as we intend them—with either babies or birds.

And although our projected lay rate was based on poultry charts and not unusual for a caged operation, maybe even a bit low, I see now it's pretty dang high for birds that would take months to fully acclimate to their new environment. Or really for pasture birds at all, chickens who simply endure more egg-withholding stress, from weather to brushes with predators. Of course, we couldn't have dreamed up a transition as traumatic as the one they'd experienced during the Myron handoff, but even so, those egg projections were lofty.

While Jason talked on, I pictured him taking out the bags of birds, throwing away thousands of potential dollars, not to mention the actual six bucks per mature hen we'd paid Myron yesterday. I swallowed this fact and tried to just listen to his voice, the phone pressed against my ear.

Jason sounded hollow, defeated.

I wasn't sure how to comfort him, as I knew each chicken lost was an emotional blow to him, as both a steward and a business-person. When anyone starts a new venture there's always a certain amount of claiming a title and growing into it, but with one's own farm, the learning is less curved and more a vertical endeavor. One day you're a guy standing in a field, the next you're the caretaker of hundreds of needy critters. If Jason had worried about this transition, he hadn't let on. Looking back, I probably hadn't created the space for him to have doubts, being so worried myself.

I made plans to meet up with him at the farm a bit later and spent that early morning shoring up the domestic front with groceries and laundry. I'd also contributed to our family's financial health by turning in another magazine article. It had been hanging over me, the way most everything had been hanging over me. I had found it difficult to write because it meant jumping off the "readying for the farm" treadmill of trips to the hardware store and afternoons stapling chicken wire under coop floors. And when I stopped, I thought. Terrifying thoughts like, *What the hell are we doing with our lives?*

I've read about people choosing wild journeys, folks who join the Peace Corps or go off grid to start a commune, individuals who don't care that they've deviated from the anticipated path or what you think of them. I am not one of those people. It has struck me more than once that I should be reading about this venture rather than writing it. But to be honest, alongside the distress of living outside social expectations, it can be exhilarating. Truly. Like we were getting away with something—a valiant, lionhearted something.

Until the next moment, when it was again terrifying and I felt the mocking shroud of grocery-cart-living loserdom descending on our household. This continual swing between these extreme poles made it difficult to shepherd words into magazine articles. Or breathe.

Before turning the minivan toward Locally Laid, I quickly threw together a care package. Jason had supposedly been eating at one of the two restaurants in Wrenshall, with humorously similar names, the Boneyard and the Brickyard, located not even a block from each other. But in actuality, Jason wasn't eating at all. When the children and I drove out we had clean clothes, snack bars, fruit, and, here I puffed up with my ingenuity, a plug-in kettle to heat water. With hot water, Jason could make himself the high-end instant coffee I picked up from Starbucks.

Coffee has always been a connection point for us, as we'd long spent lazy Sundays reading books, convening at the Church of the *New York Times*, or taking travel mugs on strolls around the lake. And now I was proud to announce that the prairie was caffeinated.

Later, when I walked him through his box of goodies, he nodded his thanks, grabbed a banana, and walked back out on the field.

"Wait!" I said, gracelessly jogging after him, the pot and dangling plug still in my hand. "Now you can make coffee out here."

"That's great, Bird, I just don't have time," he said with scarcely a backward glance as he walked toward the barn.

I stood there wondering who the hell this man was, but shook it off and moved after him. There's always so much work to do.

"I've been watering birds since sunup," Jason said. "It's incredibly slow."

He wasn't joking. The process was to fill two fifty-five-gallon drums on the back of the four-wheeler from a garden hose spigot attached to our landlady's house. It would take about forty-five minutes to top up, and then he'd drive over to the fence, getting out of the vehicle to open and relatch the two sets of metal gates, then jostle the ATV about an eighth of a mile out to the first hoop coop. From there, it was a tedious gravity-fed process via hose into the eight-gallon plastic waterers he'd gathered up in one location. Once they were full, Jason had to reposition the waterer stations, each weighing some seventy pounds.

This one task alone had taken Jason hours—and keep in mind, it was for half of the flock. The rest of our birds were due in a month.

Before we jumped in with chores, the kids and I took a few minutes to cross the pasture to where the hoop coops' doors remained shut. Myron had advised us to leave the birds inside their new structures for a day so they would identify the coop as their home and acclimate to their new surroundings. As we approached, the birds closest to us moved back, causing a ripple of movement throughout the tarped building.

We clicked our tongues gently at them, sounds that stimulated our five chickens back at home to approach, tilt their heads engagingly, and gently peck at our shoes, hoping for an apple core or some other fine treat to fall from above. But these birds were clearly different. They had none of the languid, relaxed behaviors I'd come to expect in hens but rather gave off a palpable anxiety from their staccato personalities.

We hoped feeding them would help. The sun was hot and the prairie felt exposed as we approached the feed bin. Abbie and I worked together to wiggle open its hatch—a little like freeing a bulk item at the co-op, only on an industrial scale. It was a good-sized vessel, the bin, some sixteen feet tall and holding four tons of a blend of corn, soy, and alfalfa.

I went to lift the first full pail and it quickly became a physical comedy sketch out of an old black-and-white movie. There was grunting and spilling, staggering, and a near fall. The children, not sure how to read the situation, swallowed their laughter.

You know you've been a glorified typist for most of your adult life when you try to hoist a thirty-eight-pound bucket waist high onto the back of a four-wheeler. I started filling pails only halfway. Repeating this eight times, we were then ready to drive over to the first paddock. The kids switched off at the wheel. In addition to being good driving practice, it was the closest thing to actual fun they were going to experience today. Not to say they didn't think chickens were cool, or the farm neat—in theory.

When we got to the first paddock, the first thing we did was disconnect the electric fencing. Even though the birds were not out, the fence is as much to preclude predators as to keep chickens in. That involved unplugging it from what I call the Box of Zap. It looks like a brown vintage suitcase with a futuristic solar panel built on to its smooth top. Though it might have looked harmless enough, it could deliver a swift jolt. Not the kind that'd knock you off your feet, but you'd likely jump in pain and perhaps use some creatively foul language. Once unhooked, the fencing was a harmless, pliable plastic netting on easy-to-move stakes.

In these early days, I would carefully open up a little space in the fencing to slip through. But as I got more efficient (and stronger) I learned how to step onto the fencing, caving it into the paddock and stepping off it once inside. It would reliably spring back up to place.

As I made my way in with the buckets, Abbie and Milo wandered off to the chicken infirmary, leaving me to approach the coops alone. I steadied myself with a breath and quietly begin singing as I opened the door.

"Mi-chelle, *ma belle*, these are words that go together . . ." I was shrieking before I could finish the first line.

As I'd stepped inside, there was a frenzied burst of clucking and wing flapping that cascaded into a swell of riotous movement. I fumbled with the bucket and instinctively slumped into a crouch, making myself smaller and less imposing.

"All right, all right, all right," I whispered.

I inched my way to the feed pans we were temporarily using while the girls were cooped up, causing a small kerfuffles of wings and clucking as I moved. Tentatively, quietly, I shook out feed. There had been some left in the pan, but the sight and smell of a fresh meal perked a few up. Some even let out a bit more of a happy *ca-caw* noise.

"There you go. See, Lucie's not so bad, no," I prattled on to hold my nerve.

As I slunk out, back first, into the full sun, I sat on the grass and exhaled a ragged breath.

The kids were heading my way now. Abbie's face was red under her fashion cap, built to frame her face rather than shield it, while Milo slumped along after her, generally carrying the look of the oppressed. Jason, our field expert, didn't have the mental bandwidth to engage with them like he typically would. He likes to talk through things as he works, be a hands-on teacher to the kids, but today he was too busy being schooled himself.

I sighed, pushing out thoughts of what a great beach day this would be. I couldn't allow myself to think that way anymore; this was our life now. We were making a choice away from a life of steady paychecks, paid vacations, and weekends off for this, the unpredictability of the pasture.

When the children and I returned the next day, we saw Jason in the distance heading up the rise to the coops. Out of our six structures, each a good 250 feet apart, three were filled with these new birds awaiting their first introduction to pasture. After they had spent a full twenty-four hours getting to know their coop, it was time to expand their world by a few hundred feet of green prairie.

Walking up the slight incline to where the coops were, it would be hard to overstate how beautiful the pasture was, thick with clover and tiny flowers. I half expected Julie Andrews to leap out in her peasant outfit and belt out "The Hills Are Alive." By the time we got there, Jason was ready to open the door on the first coop, freeing the chickens into more space than they'd ever enjoyed. From being raised and transported in confinement, this, right then, was the point of the whole saga, theirs and ours.

As Jason unlatched the screen door, opening it wide into that gorgeous summer morning, the chickens backed up. They wanted no part of it. Absolutely none. It took us over an hour to shoo them all, crouched low to shepherd them out the door and to catch those trying to sneak back in. Over an hour later, three coops were emptied and all of the surviving hens (an approximate 775 surviving of our original 900) were out, doors shut behind them.

They stared at us.

Our home flock of five chickens are mighty foragers, and when they strut out in the morning, they're on the job. They score and scratch the earth with their reptilian claws, then hop their whole chicken bodies back to survey what they've turned up. Pivoting their heads a full ninety degrees to get a side eye to the ground, they're merciless with bugs and can scoop one up with blurring mechanical speed.

They're peckers, all right.

But the farm flock in front of us was, well, stupefied. They didn't seem to know what to do. Our entire pasture-raised concept was predicated on their grazing, but they were looking at us as if awaiting instructions.

Jason and I went quiet. We stood there, watching the chickens. And the chickens stood there, watching us, until one hen put her head to the grass and, as though in painful super-slow motion, she . . . pecked.

Gradually, the behavior rippled through the flock and the birds commenced ungainly foraging. Over the day, as we slung feed and filled the waterers, their skills developed. By afternoon, some flies had gathered on the sunny side of the coop, warming themselves against the tarp. A bird or two approached and clumsily snapped at the insects. Eventually, they caught a few.

The children and I stood in the field with the birds, watching them figure it out. If we leaned in close enough, I felt we might see the virtual lightning bolt snap through their little chicken cerebrums, like they were remembering something they didn't even know they knew.

After we'd worked our way through the daylong chores, it was getting late and I needed to get the kids home. But I wanted to see the Poultry Butlers in action. I had a soft spot for these hilariously named solar-powered products. They're hen doors that sense the sunrise and open—letting birds out. After chickens follow their natural instinct to seek shelter at dusk, the doors shut behind them. This would button up the girls' house for the night. The combination of ridiculousness and practicality made me smile. Actually, at $170 each, these little doors even made it into the business plan. Because of them, we felt we could live away from the farm and know that the chickens would be protected against predators.

"The sun is really dropping now," I'd observed, shifting weight from one hiking boot to the other.

But these hens were not going inside. They stood there, eyeing us.

"Huh," Jason said as he rubbed his hands down on his dirty jeans.

The children stood quietly watching the chickens half-peck at grass in the fast-fading light.

Nothing happened.

And then it snapped into place . . . for me.

"Jason, they're not going to go in."

"Just give them a chance," he said, slightly irritated, though not really with me.

"No, see . . . they're not responding to night, because they've never actually experienced day before—ever."

His eyes widened and we turned our attention back to these birds that could not make sense of a world without fluorescents. I realized something even more unnerving: we were going to have to get hundreds of hens into their coops.

I muttered an obscenity. One of the really bad ones. You may think, "How fast can a flightless bird be?" Consider the *Rocky* sequel. In it the boxer hones his speed and agility by trying to catch a chicken. It takes many segments of grueling montage before Mr. Balboa can overtake the sprinting hen. Of course, the crappy parts of my life refused to compress into a neat segment scored with inspirational music. I suffered every minute of the next hour, bent over attempting to rush birds, while getting a fluster of wings in my face.

Many hopped back out, until we had the good idea to station Milo at the door to act as sentry.

When the last bird was in, I leaned against the coop and felt all motivation drain from my muscles. We'd never be able to trust the chickens to put themselves to bed, nor, I was sure, could we count on them to come out in the morning. There would be long days here or we'd have to hire locals to do these tasks. Either way, this was not good news for Locally Laid. I was ready to go home and indulge in my second self-pity bath of the week. Then I looked inside.

"Jason," I said with new alertness, "they're all on the floor."

It was completely dark now as Jason flipped open his cell phone and illuminated the hundreds of hens' eyes, which reflected red and green. Not one bird was on a roost.

Now if it was instinct that drove a bird to shelter at dark, it was that same sense of self-preservation that would tell her to seek height.

A roost is nothing more than a highly placed stick or a branch that a bird clutches her feet around when at rest. It's likely a throwback to their jungle heritage when chickens escaped to the trees for night safety. The practice still serves them well here in the heartland. It allows an opportunity to air tail feathers, tamp down bacteria, and perch close to each other for warmth and protection and out of their own fecal matter. (Chicken shit happens, but no one wants to bed down in it.) More important— it's a real estate issue. We'd based the footprint of the shelter on vertical use of space. To say this another way, it was like we built a high-rise apartment and everyone was racing to live in the ground-floor unit. The danger of suffocation was real.

Jason sent us home and decided to wait a few minutes until the birds were fully asleep. Chickens sleep hard. Whoever came up with the saying "dead to the world" probably first uttered it in a henhouse at night. Whereas walking among hens in the day will cause a flustery ruckus, the night coop is peaceful with only the occasional *coo* sound emitting from sleeping birds, heads tucked so far under their wings they hardly look like they have heads.

That was when Jason walked in wearing a headlamp. He gently picked up each sleep-immobilized chicken and hand-placed them on their roosts, one by one by one. The birds instinctually wrapped their chicken toes around the stick despite their heavy-lidded slumber.

Because it was summer, the sun didn't fully drop until nearly ten p.m., meaning Jason was either sleeping on the cot in the egg-washing building or coming home around midnight, only to be out the door again well before six a.m.—his new wake-up call.

Although it took over two weeks, the birds did come to roost on their own as Jason managed to rewire their walnut-sized brains back to their instinctual state. (My friend Deb joked that each bird likely woke up in the morning thinking, "It happened again!" And looked over to her nearest roost buddy and said, "You too, Delores?")

But our hens weren't the only ones with new realities to adjust to. I began to see that our lives had shifted as well. Jason was now a farmer, and I was a married single parent.

Chapter 8

I am not a risk taker by nature, and there's a case to be made that I'm just a plain weenie. In the section of my heart reserved for stout entrepreneurism sits a shirking pinto bean or maybe an eraser head. And it's probably, counterintuitively, because I came from a business-owning family.

What I remember about my parents' building construction venture was that it employed nearly a dozen crewmembers and was both successful and filled with stressors—or as I call it, succestressful. In my memory, I see my father shoveling dinner into his mouth at the kitchen table while talking on a telephone stretched to the end of its curly cord. It was often a conversation steeped with worry about subcontractors or a foundation not setting up properly due to a cold snap. He'd often get out of bed at night to check a job site.

My father had very little time, heavy thoughts, and ulcer trouble. My mother worked nonstop to keep a lovely home and then, after dinner was put away, hunched over wide ledgers and a metal-handled check embossing machine.

It was a few weeks after the first batch of chickens arrived that I went to see my parents, now retired, in Maine. When I visit my home state, I almost always finagle a writing gig to at least partially morph my annual visit into a business trip. The most amenable has been *Cabin Life,* a "lifestyle" magazine featuring glossy spreads of luxuriant second homes on breathtaking properties—real estate pornography.

I took my father along on that day's adventure to Cundy Harbor, a seaside fishing village so obscure that GPS simply gave up. The Maine coast is tricky that way, a collection of crooked and curved land fingers constantly washed by the Atlantic. We had just emerged from the hairiest of this driving, my seventy-eight-year-old dad at the wheel. Our chatter had been about the restored Swedish-style cabin we'd just toured when he turned his head and asked what had clearly been on his mind for months.

"Does Jason have any idea what it means to take care of animals?"

The directness of the question pulled my stomach to my throat, rolling an organ or two into a knot along the way. This, of course, was a legitimate inquiry. And as I'd opened my mouth to say I really don't know what, he continued.

"And the farm, so far away from your house—on rented land—is that a good idea?" Clearly, he had concerns, valid ones, and I was stuck in the untenable position of defending a venture I was still trying to make friends with myself.

What I also found out on that long drive was that my father had worked with chickens when he was a teen. He'd been a farmhand at an egg-producing farm not even half a mile from where I grew up. My father never mentioned it before. At least that I can recall.

Driving by the same field, he pointed out where the structures had been, a typical late 1940s floor operation. Today this would be called cage-free—chickens stayed inside, freely roaming in the barn. This one housed several thousand birds on three levels.

When my father arrived in the morning, his first job wasn't to collect eggs; rather he gathered up the chickens that had died overnight. There were several a day, which he'd take out to a field and bury.

Once back inside, my father did all the feeding and watering. Just a skinny kid, he'd swing hundred-pound bags of feed that weighed as much as he did over his shoulder and carry them up a ladder to the third floor. All this he performed for some thirty cents an hour. No wonder he was dubious on the topic of egg-related work.

Only after my father quit, when he was old enough to go into the carpentry trade, did the farmer he worked for install a freight elevator to motor the heavy feed bags to the top level. Nearly seventy years later, this fact still irritates my dad.

Looking through this lens, I see how bizarre this new farm venture was to my father. Especially as he and my mother had ensured my brother and I got educations, allowing us opportunities they never had. And here I was, making an about-face from progress.

I couldn't shake my father's chicken history.

Historically, agricultural operations flecked across not only Maine but the entire nation. Less than a hundred years ago, small to midsized food producers numbered nearly six million across America, the kind of operations that frankly would've been great mentor farms to a fledging operation like Locally Laid and could have easily hooked us up with a couple thousand birds.

Their vanishing made me ask—what happened? And this is where I attempt to cram seventy-five years' worth of history into about seven hundred words.

The more I read about this subject, the more I now see that our nation's entire food system pivoted on World War II. Picture the stern couple in Grant Wood's *American Gothic* painting. You've seen it. Perhaps not the original oil rendering hanging at the Art Institute of Chicago, but surely you're familiar with the much-parodied work of a balding, middle-aged farmer gripping a pitchfork, along with his dour, spinster daughter. The portrait was modeled by the painter's sister, Nan, and his dentist, Byron, but humor me for a moment and let's pretend they're Mr. and Mrs. Gothic, prewar farmers.

You can see by Nan's and Byron's serious posture and the decisive pitchfork placement that they were living in the tough times of the Depression and the Dust Bowl. But surviving that, and coming into the 1940s, they'd likely have had a tractor to break up the land and help comb out the harvest. Despite this mechanization, there'd also have been farmworkers, many of them. Weed control was still largely a task done literally by hand.

Then Germany invaded Poland, the Japanese bombed Pearl Harbor, and, with America's entry into war, most of the Gothics' crew would have been drafted or enlisted, caught up in the zeitgeist of patriotism. In fact, the nation's farm population declined by 17 percent, just as farms were expected to produce more. Food was needed at home, for troops, and even for starving Allied countries.

The only way to feed everyone was to leverage the labor force left behind—and produce more crops with fewer people. Ford-Ferguson saw an opening in the market and began producing a line of easier-to-handle tractors, advertising that even "children and old people" could operate them. It was an all-hands operation.

And here's a neat fact: our Gothic farm couple at this time would have had money (perhaps tucked into Nan's colonial-print apron) as wartime farm incomes nearly tripled. And when peace finally arrived, many used their profits as down payments on bigger, better, more efficient implements. They needed them to replace the many hands not returning to the farm, opting instead to pick up pencils the GI Bill bought for them. However, what did come home from war was the chemicals used to fight it. DDT and nitrate fertilizers, now made in former ordnance factories (often sharing some of the same ingredient lists), were promoted as yield-boosting, labor-saving options.

So, for a while, farmers enjoyed a bit of an agricultural bender. (I like to picture Byron with a noisemaker tucked in his overalls and Nan wearing a jaunty party hat.) And just like an infomercial, it somehow got better. The Marshall Plan, also known as the European Recovery Program, had the United States buying billions of dollars' worth of produce to ship to rebuilding Europe.

Of course, binges never end pretty.

Eventually, Europe and Asia bent their swords back into plowshares to once again produce their own harvests. This effectively closed a large market for U.S. food. Add to that, crop producers were realizing a bit of a chemical dependency. Increasing amounts of nitrogen were being tilled into fields, with fewer returns.

So, while the fertilizer companies bathed in money, as farmers scrambled to increase yield to pay for those new tractors and combines and increased acres, farms went under. Push that famous house with the Gothic window right out from behind Nan and Byron.

And you thought they looked austere before.

The truly sad part is this: it wasn't needed. The corn, that is. The Gothics and thousands of farmers were killing themselves (and their soil) to crank out a crop we had in abundant surplus. This leaves us with the obvious question: *For God's sake, WHY?*

The answer was politics. Oversupply played a role in international relations, serving as a double-headed carrot-and-stick tool of diplomacy. Food aid became a reward to nations that served the interests of the United States and conversely could be withheld or used to undersell those who did not. What wasn't wielded internationally was sold cheaply to domestic food processors, becoming the ubiquitous ingredient high-fructose corn syrup. Looking at our current obesity rates and other health concerns, we see how that turned out.

Things got more precarious when U.S. Department of Agriculture secretary Earl Butz—a man infamous for telling farmers to "Get big or get out"—dismantled Franklin Roosevelt's New Deal programs that had regulated commodity prices. With those protections gone, farmers were exposed to the caprice of the marketplace. Oversupply in the 1980s hit critical levels and prices bottomed out just as I remember Walter Cronkite on TV soberly telling the nation about all-time-high interest rates. Throw in a couple of droughts (and your Farm Aid T-shirt) and 1987 became the year that the U.S. Department of Agriculture (USDA) recorded the highest number of farm bankruptcies in its history. Ever.

And that was why we were stuck getting chickens from the likes of Myron.

Chapter 9

It was around then that I began to suspect LoLa (our shorthand for Locally Laid and also the name of every chicken on the farm) was a thief, scalawagging off with my hammer, my drill, my sleeping bag, my flashlight, and, of course, my husband. Even my daughter had been recently relocated to the farm.

Abbie was a natural on the pasture, confidently striding over the flexible fencing to identify the smaller birds who'd been bullied away from the feeders. These were essentially long troughs made from PVC gutters drilled to two-by-four boards, a system Jason had seen when visiting a sustainable farm in Maryland early that spring.

What we later learned was that the setup was terribly wasteful and beyond that, bad for bird harmony. One of the problems was that the gutters had no lip—a wedge of plastic to keep the hens from tossing out lots and lots of expensive feed on the ground as they picked through it looking for the good stuff, bits of sweet corn.

The other problem was that these gutter-style feeders led to competition. Because they're essentially open troughs, they couldn't refill themselves like an auto-feeder, a simple container above that would gravity-pour as it emptied. Thus the chickens became fierce about getting their fair share—and then some. Big chickens pecked away smaller chickens and literally ate their dinner, until those big birds morphed into visibly dominant hens and the smaller, now scrawny birds scarcely ate anything at all. Not nearly the four ounces of daily feed needed to produce an egg.

Abbie created a place for these runts by nudging more aggressive chickens with her tiny hiking boot. She was no longer squeamish about picking up injured hens, cleaning out the wounds, and then slathering them with ointment, sometimes butterfly-taping them shut with an adhesive medical bandage. I liked to watch as she plied her careful treatment, gently talking them through it. At age eleven, she looked like she'd been doing this all of her preteen life.

I remember one particular day in late June. The chickens still needed to be ushered out of the coops in the morning and marshaled back in and roosted at night. The earliest layers were beginning to give their very first pullet eggs, usually a robin-sized offering. There were not yet enough eggs to start official processing and sales, just the very start.

Abbie opened a chicken OB-GYN practice of sorts, helping egg-bound hens with what amounts to ovum constipation. There had been a spate of such troubled birds. They'd sit in the nesting box, hunched up, feathers ruffled, and display their discomfort by contracting their feathery back ends and wincing—a facial expression that crossed human/avian lines.

After consulting Dr. Google, Jason showed Abbie how to identify the hard knot of an unpassed egg with her fingertips. Then, using a gentle pushing motion toward the bird's vent, she attempted to stimulate the shell gland, the equivalent of a poultry uterus, to push out the lodged egg. Careful not to apply heavy pressure that might break the shell and risk infection, she'd rub down the chickens. It was not unlike the tummy massage Jason and I did once on baby Abbie, swirling our hands around her abdomen in an attempt to relieve colic.

This seemed to work about as well on a crying infant as it did on an egg-bound hen, which is to say not very. When a stuck egg refused to be dislodged, the father-daughter team would sterilize a tool, say a slender screwdriver, and try to break the stuck egg with one clean puncture before the hen's straining caused prolapse.

And prolapse is bad business. Very bad business.

According to *The Chicken Health Handbook*, prolapse is when a hen's oviduct is pushed inside out and protrudes from its vent. Colloquially, it's known as blowout. It presents as a bright pink fleshy bulge under a bird's tail feathers and makes me grimace with labor empathy.

When Abbie would find a blown-out bird, she knew to immediately separate her from the rest of the flock, or truly the worst would happen. The angry color of the exposed organ would attract pecking. (You recall that chickens, omnivores through

and through, are drawn to peck anything bloodred.) This pecking would lead to tugging, which would eventually haul out the
bird's intestines. This meant more exposed flesh, more blood,
and more pecking until the chicken died under the sharp beaks
of her sisters. No matter how sweet a hen was, the pecking order
and its mandate to keep the flock healthy and strong would be
upheld. The laws of nature allowed for no weakness.

However, if that injured bird was put into her own space, the
wounded area cleaned, and the organ gently pushed back with a
gloved hand, she might make it. Maybe. We'd never seen prolapse
with our home flock, but it can be a symptom of a number of
things from feed blend to early maturation. One reason, we'd
been told, might have been that the type of calcium mixed into
their feed was too fine for proper absorption. It's a nutrient needed
to stimulate the pushing action that safely pops the egg out the
vent. I'd heard Jason on the phone changing to a more expensive
vitamin mix with a coarser form of the mineral. At this moment,
we'd have paid anything to avoid these gruesome deaths.

I'd also read that prolapse was a problem in underweight
birds, like the ones we'd received from Myron. The difficulty is
that the hens would sexually mature before their bodies were
physically ready to handle it. I wish we could have fixed this with
a simple outlay of cash.

Abbie had also been out rotating pasture with Jason, no easy
task. What made this a formidable chore was twofold. First, while
the flexible fence is light, hauling a good hundred feet of it isn't.
It gets unwieldy fast, especially as it's dragged over uneven ter-
rain or through longer grasses. Second, unless you really think

through your strategy, wily birds will spot a gap as you wriggle the fence along—and bolt.

These were our early days, when we were still developing our best-practice techniques. Now we know to walk behind the hens, ushering them along, moving fencing behind them as we create a new, fresh "petal" of enclosed paddock off the stationary hoop coop. But back then we were stumble-tripping along. And by that, I'm saying we chased a lot of chickens.

When one accidentally released hens, it meant jogging over to the four-wheeler, grabbing one of the fishing nets, and chasing the birds through the open prairie. Jason was best at this, returning to his Minnesota hockey roots, treating it like another puck drill. The time this exercise would take depended greatly on your (and your birds') athleticism; it could mean a minute's romp or up to ten before a bird was swooped up, multiplied by the number of times you let her escape.

Fortunately, this labor-heavy effort of pasture rotation holds an immediate payoff. Once the chickens are shown their fresh field, it's nothing short of poultry jubilation. Seeing the verdant grasses, they're in full pursuit. They tuck their wings tight against their bodies, canter forward, and rapidly pick up one foot and then the other for intense speed waddling. The bobbing articulation of a chicken's momentum, jutting her head forward, then waiting for her body to run underneath it again and again and again, well—that's entertainment.

Sometime between that morning's feed haul, hemming in a fresh paddock, and playing Prairie Bird, MD, Abbie didn't just get hungry; she was ravenous. The granola bar and banana Jason had brought for her wouldn't come close to satisfying that. This called for a trip to town.

And that was how Abbie traumatized a small child.

Our daughter found herself in a sandwich shop wearing her Chester Bowl sweatshirt, not only boasting of our neighborhood ski hill but now also splotched with a terrific bloodstain and patches of feathers. Of all our family members, Abbie is the best hand washer, and she tried her best to clean up the spatters of farm life from her arms in the small public restroom. But there was nothing to do about the garment.

"There was this little girl there that I'm pretty sure I've mentally scarred," Abbie told me later. "She just stared at me with giant eyes, like I was some monster that she couldn't let out of her sight."

I'd have felt terrible about this statement if Abbie hadn't shown a slight smugness with her grin, a little light in her eyes. There was an underlying tough pride to her that I quietly loved.

Milo, however, was not as comfortable out at the farm. Like, not at all. We share a slight coordination disorder called dyspraxia, which makes us a bit grace impaired. In short, we're genetically clumsy—really, a cruel joke for a kid whose family now owned an egg farm. He'd accidentally tip feeders and trip up fencing, releasing dozens of birds. This caused him to flush with frustrated shame. We decided to limit his time there, at least until he was a bit older. I had to constantly remind myself that he was still a little guy, only nine.

Milo's farm role was taking over the backyard chicken duties at our house in Duluth. He was responsible for the five home hens and, in return, he sold their wares—for personal profit. He's always been a boy of unusual interests and for as long as I can remember, desperate to be an entrepreneur. In first grade, he'd cried about having to go to school, where he wasn't learning anything that would help him become "a captain of industry." Though his end goals seemed to skew more toward robber baron, I didn't dispute it.

I'd like to tell you that Milo was modeling the venture of Locally Laid, or even me with my decade of freelance writing (though my "gal with laptop will 'wordy pole–dance' for you" business setup wasn't that complex). But in reality, Milo had been inspired by documentaries on the Rockefellers and Carnegies and the reality television show *Shark Tank*.

This spurred my son to start his own budding egg-pire. Wanting to drive sales, he had an idea to put something "value added" into his cartons, like cereal companies do with cheap toys. However, when I explained that he'd have to buy the object and that would come out of his profits, he blanched. He figured he could create something instead. Milo wrote out knock-knock jokes from library books and clipped comics from the newspaper, tucking them into each carton of his home-laid eggs.

Thus, *Milo's Yolks and Jokes* was born.

The part he liked most (and was best at) was the sales. He developed a flyer (copying each by hand), knocked on neighbors' doors, and made handshake deals. Later, Milo expanded his income stream by adding advertisements to the top and inside of his recycled egg carton lids. Friends Andy and Katy, who run a local acting theater, graciously purchased a few announcements for upcoming shows. Abbie drew the ads (Milo paid her a small sum) and stuck them to the cartons heading out to his customers. No one complained about the eggs and spam.

While my husband all but lived on the farmland, dragging himself home after dark for a shower and quick sleep before heading back out, my prairie time was becoming limited to weekends and the odd day during the week. That was because one of my gigs at an ad agency had grown to the point that I was working in its office.

And I loved it. Going to Swim Creative was everything that the farm wasn't. In addition to being dry and temperature controlled, it was in downtown Duluth, where one enjoyed the bustle of a retro Main Street and the sun sparkles bouncing off Lake Superior. Plus, it smelled nice.

When typing on my Mac on my agency-provided Ikea desk, it was easy to forget that I was an egg proprietress at all. Until I'd suddenly remember and the thought would dump ice crystals into my bloodstream. Sometimes, it'd drive me to hide in the bathroom for a few minutes, collecting myself until I could dive into the next project and forget again.

In addition to liking the work, I was meeting people. For years, I'd spent my workdays primarily alone, and I can attest that the living room couch does not afford you many new acquaintances beyond the persistent gentleman with his issues of *Watchtower* magazine.

While I enjoyed all my co-workers, I was particularly fond of Beau Walsh, the guy with the desk next to me. He was a designer hired by the agency right out of his internship and had recently enjoyed his first legal beer. We worked on videos together, and I could tell he had an innate talent for storytelling. Together we made a nice team. Later, this working relationship would become very important to Locally Laid, but back then I was just happy to have Beau as a sweet pal.

Of course, the ad agency wasn't my only gig.

Like Jason, I'd do early-morning farm chores, but mine were on my Mac. I was building a brand and it started with the name.

"I'm thinking Amundsen Farms," Jason had said, repeatedly testing its echolalia, sitting at his antique secretary. This was back in our beautiful BC days: Before Chickens.

"Amundsen Farms?" I made a look of distaste. "You can't have a name that's not instinctively spelled."

The *s-E-n* is an unexpected trip-up at the end of Jason's Norwegian surname, and frankly the whole thing looks harder to say than it is. It's pronounced *Ah-mun-son* (with a silent *d*), like the Antarctic explorer, but if telemarketers are any gauge, it's tough on the uninitiated.

After a bit of conversation, I'd convinced Jason that the name should showcase how we would stand apart in the egg case, a differentiator as the marketers would say. That was when we landed on a moniker highlighting our regionality: Locally Laid. And though I was admittedly shy of its double entendre, it truly is a perfect one with a local focus.

Back then I knew only that local food and its devoted follow-ers, locavores, were, you know . . . a thing. A growing, popular, cool thing, for sure, and one doing good in the world, though I couldn't exactly articulate what those good things were. Clearly, there was the obvious point of freshness, but beyond that, things got fuzzy.

That was before I'd heard of the newish term *food miles*. It's a form of culinary accounting, adding up all the travel on long-haul trucks and plane hops a product takes to get from farm field to processing plant to your town. The Worldwatch Institute ran the numbers and figures that most things we pick up from the grocery store travel between 1,500 and 2,500 miles before they slap down on our kitchen counters. And that seemed like an awful big side of diesel with our breakfast. I started writing about it.

Later, I would understand how much more important local eating is, not just for fuel consumption but for regional economies and sidestepping the problems of food production concentration, such as the overfarming of land and feedlot waste. But food miles was a good start for my researching and writing.

I was beginning to see myself as a pixel farmer writing about our cutting-edge retro practices.

Sometimes it was an urgent task, like when I wrote and rewrote the copy for the egg cartons we needed to get printed for upcoming sales—only to be rejected again by the State of Minnesota for word choice, placement, or size.

As there is no official federal language for pasture-raised eggs, it was a bit like a segment of *The Price Is Right*. I'd arrange all my words, run back to pull a cartoonishly large handle, and get the buzzer—though it was often unclear exactly what was wrong. And I wouldn't know until I'd reordered all the words over again. When I'd gotten that straightened out, there were always formatting issues with our carton printer.

Also, I kept the farm website and the Locally Laid blog. Serving as part diary, part therapy, part customer interaction, LoLa's social media already had a small following of fellow dreamers and the ag-curious. I had a theory that they read along because I didn't completely sugarcoat it. I told them the things that went wrong—like when we got a vehicle stuck or shocked ourselves on the electric fencing or whatever chicken-chasing nonsense the day served up. It's a balance. I wanted to keep people rooting for us like the *Little Chicken That Could* but not push us over into the likes of *Jackass 3*.

It was sometimes difficult to write about these often humbling days out at Locally Laid, but it also offered me the chance to reframe the experiences. As a child, I watched reruns of *The Carol Burnett Show,* whose star famously coined the secret formula for funny: Comedy is tragedy, plus time. Writing our farm antics in real time compelled me to speed up the process and expose that much-needed humor right away.

Most followers seemed to appreciate the kind of forthrightness you're unlikely to see with, say, a Kraft or even other start-ups. In literary terms, it made me a reliable narrator, and because I told people the bad, it made the good I shared all the more believable. After just a few months online, we already had several hundred followers on Facebook (even the woman who did not eat eggs). It was a group that clamored for chicken pictures so ardently I've come to refer to these snapshots I'd post as Softcore Poultry.

While Jason mildly appreciated the "atta-boys" we'd get, I knew he'd do the work without any outside encouragement. I'm more shallow and take people's sincere interest as much-needed affirmation that our farming upstream was worth all we endured.

But it was becoming clear that people were starting to groove on our mission, and that was when it became clear to me that Locally Laid had one.

I often don't really know what's right in front of me until I start writing about it. That was how I came to see that our ag business had a bigger guiding force than "let's sell enough tasty farm eggs to avoid eviction." It was gelling into an actual ethos, practically a life of its own that would come to embody our little logo chicken, LoLa.

That logo, which perfectly captured the personality of our farm, is the work of Matt Olin. We had paid out some modest sums to other designers, auditioning concepts, but when Matt showed us his design we knew we'd found our girl. Her legs sprout out of the lowercase *L*s and her body swoops around, gray and speckled, to her backward-facing head and orange triangle beak. She's also just laid a beautiful fresh egg into the crux of the "a." As far as chickens go, she's perfect.

I was beginning to feel like my graduate student friends who wrote fiction and told me that their characters mysteriously took on a life of their own, dictating the story. While I wasn't ready to say that LoLa ran the show, she did seem to have opinions.

As I started to parse and play with the language of Locally Laid, themes started to emerge. There was a locavore element, an animal welfare bent, and something solid for the local business enthusiast. Once we started selling commercially, an ardent foodie factor emerged, and there were always the folks who just like chickens. Correction: people who REVERE chickens, like just this side of poultry obsessed and needing help.

It would take me a good two years of writing and learning to be able to knead all that we do into a true vocation, a calling, our mission.

ACT 3

Egged

SLANG: a foul prank

Chapter 10

By July, about three weeks after our first chickens came, the second batch of nine hundred hens arrived. And, Myron being Myron, we got little notice. The birds were sprung upon us on a weekday while the children and I were still on the East Coast on writing assignment and visiting family.

"Don't worry, we'll be okay," Jason said over the phone.

His use of "we" was meant mostly in the royal sense. Though I was doing my part of writing for cash and sharing in chicken chores on the weekend, I felt like I'd deserted Jason. He was unmoored in the sea of details. I'd find his lists of things that needed to get done immediately, if not sooner. They'd be written on scraps of paper torn from poultry vitamin sacks or peeled Gatorade labels. I knew my contribution as home fires manager was both important and essential, but I felt that Jason and I no longer existed in that couple co-space we'd long shared, where it was him and me against the world, especially now.

We were only working at the farm together maybe one day

a week, usually in separate paddocks, and had no time for recreation. All the things I loved—hiking, biking, and going to the beach—were exactly what Jason didn't want to do with what little time he wasn't farming. He deserved to hide indoors, in front of Netflix, fleeing both the real and outdoor world.

Not being there for the next batch of birds, what we called at home The Second Coming, just added to my sense of displacement. I suspected this bothered Jason, too, but there was no time to think about it, as another slapdash delivery of hens was happening.

While the knowledge accrued from that first chicken delivery helped mitigate the problems of the second, it was, according to Jason, nothing short of heinous.

He sent me a text at eleven p.m.: *Many casualties.*

It was nearly as many as before. And this batch also required Jason's pasture- and roost-training regimen, meaning another three weeks of sleeping at the farm. Except there wasn't as much room for Jason in the wooden egg-processing house now that our egg-washing machine, the retro-laborious, circa 1954 Aqua-Magic V, had been plumbed in by our friend Rod.

The old machine, found in a barn, had been refurbished by none other than Myron. I've got to give the man credit, he was good at rebuilding vintage machinery. Our washer was a long, fairly narrow metal device sitting atop metal legs and befitting of the word *contraption*. But there was no time to play with it much then.

The daily watering and feeding chores were now taking some five hours a day. Plus, the chickens were starting to get crafty. Slowly becoming good foragers, running after crickets and mowing down the field grass to a nub before being moved onto fresh pasture, they had figured out what they liked best. And that was corn. When they saw the four-wheeler, they'd run toward it, knowing there were feed buckets in the back. They'd even figured out that if they worked *en masse*, they could push over the flexible fence, pulling its stakes right out of the ground, and escape. This meant more chasing of birds, wielding our fishing nets like angry villagers forming a mob. It was a huge loss of both time and pride. It got to the point that Jason would walk all the way across the field carrying two heavy buckets of feed to serve them an appetizer so they'd be distracted when he drove up with the rest of their ration.

"I'm not that bright, but I'm happy to say I can outwit a chicken," he said.

This farm physicality was starting to wear on Jason's body, and there always seemed to be some unexpected daily bonus work requiring immediate attention: fixing a shorted-out fence, rewiring the Poultry Butlers again, or changing a flat on the four-wheeler. He was now seeing a chiropractor and sported a terrible rash on his back with painful blisters caused by a virus known as gladiatorum, named after another fun-loving, hardworking lot: Roman gladiators. Likely, Jason had picked up the virus as a sport-playing kid and had a localized case that he'd never noticed. But now as he stressed his body, it was the perfect storm for a massive flare-up that also included headache, sore throat, and, as we learned, diarrhea.

"Hey, how did the visit with the restaurant people go today?" I'd phoned Jason from my parents' driveway.

In addition to Jason getting new birds and being short-handed, he'd long promised a farm tour to a local restaurant, one that went through eggs by the thousands every month. The Duluth Grill was, back then, one of the few places in our region making the effort to buy from local farmers. That they were considering using Locally Laid, even partially, once our farm was in full production and state inspected, was a big deal.

Despite his weariness, Jason giggled into his cell. "Oh, Lu, I don't know if I have a bug or I'm just stressed, but when they were here I had a terrible case of the shi—"

"Backdoor trots?" I offered over his slang.

"Oh, that doesn't even begin to cover it," he replied.

I winced.

Apparently, Jason had been showing owner Tom Hanson, his son Louie, and other employees our prairie setup when he'd urgently excused himself. With no hope of making it the hundreds of feet to our Porta-Potty back by the barn, Jason relieved himself in a less-than-private location just behind a coop.

"Chickens on the other side of the fence were coming up to stare at me," he laughed.

When I returned, the some sixteen-hundred-ish surviving chickens were now on pasture—and Jason was elusive as Bigfoot. There was taunting evidence of his existence—foul-smelling laundry and the occasional Jason-shaped lump on the couch—but not much that anyone could claim as proof of life.

When he did come home, it was usually after dark, when he'd hork down food, any food, and hit REM sleep before he could even set his plate on the coffee table right in front of him.

More than once I'd taken silverware off his rhythmically rising chest as he snored openmouthed on the living room floor. Walking upstairs to our room was simply too large a feat, and we rarely shared a bed. I'd joke with my best friends that "ain't nobody getting laid at Locally Laid," which was funny, mostly.

The truth was, I missed him in every way possible.

One early morning in late July, as we were reviewing our kid troop movements and social calendar during a predawn roundup, Jason interrupted.

"Oh, the inspector comes today." He said this calmly, without angst or emotion, and took a long pull on his coffee.

I raised an eye from the calendar in my lap. The term *inspector* just sounded ominous. This wasn't a person tasked to issue our participation medal. "No worries, Bird, it's a good thing," he said. "The hens are finally starting to drop eggs, and we need this inspection checked off so we can get cartons in stores."

The chickens were laying later than our business plan had stated (apparently they hadn't read it), but given the condition in which they arrived we were happy they were laying at all.

"Are we ready?" I asked.

Jason was confident. "The inspection will be good, been ready for weeks," he said grabbing a dirty binder with the word *RULES* scrawled in his untidy hand on the front cover. He opened it to a printout of all the regulations, with sections highlighted and notes penned in the margins. There were checks and stars and underlining next to dirt smudges and mystery stains shaped like thumbs. I'm not sure it qualified as a living document, but it sure had seen a lot of life.

"These are the rules, and as I met each one, I checked it here." He pointed to a column of small *X*s with rough fingers that just a few weeks ago had been supple and pink. "And if I have a particular question, I've written it here," he said, sliding a finger over to the far margin on the right to small clots of scribbles. "Then I called the state and got it clarified, writing their response here."

Pride washed over me and I smiled.

Jason stood, snapped the formerly white binder shut with a snap, tucked it under his arm, and pulled me into a strong embrace.

"Gotta run, my ladies need me." He hammed it up a bit and for that split second he was truly present. I could see that, although he was fatigued, he was also incandescently happy as only a person moving forward on a grand project can be.

"It's all problem-solving, Lu, that's all farming is," he said. And it seemed he was enjoying the puzzle.

He issued me the standard married-people lip peck before grabbing his coffee and walking out the door. I opened my mouth to nag about his taking our good mugs to the farm, never to return, but I stopped myself. Sometimes the price of a happy marriage is lost stoneware.

Later, the children and I were in our backyard garden. It was a quintessential summer night in Duluth with the Free-B-Q going (that's the barbecue we found a few years back with a *free* sign on it), grilling bratwurst and mushrooms alongside long strips of zucchini and squash.

Despite the light sky, I felt time elapsing and contemplated how long to hold dinner for Jason. The brats were starting to blacken, the vegetable spears were wrinkled, and I wanted time to clean up before the outdoor movie.

I elected to feed the children. While I wasn't eating yet, I sat with them on the patio furniture to join in our "Gratefuls." That's when we all hold hands, close our eyes for a second, and literally intone the word *grateful*. It's part prayer, part intention, and, like Locally Laid, it's a phrase that can be taken any way one wants.

As they ate, I packed our beach throw into the big blue Ikea bag along with some lumpy pillows for the evening ahead at Movies in the Park. Fridays in July and August, the Downtown Council inflates a giant screen right by chilly Lake Superior and hundreds of people gather in the dark, wrapped in blankets, sitting on lawn chairs. Honestly, it's magical.

Hearing the old Accord pinching the gravel in the alley, I popped up toward our driveway. I had a Grain Belt Premium in hand for Jason, brewed in our old Minneapolis neighborhood and a loose Friday tradition. Even though we both worked seven days a week now, it was still the start of the weekend and that should mean something.

But as he got out of the car, I knew something was wrong. Very wrong. He was pale underneath his farmer's tan and though he was giving me a little smile, there was no mirth in it. Walking toward me, it looked like only half his bones were willing to bear any weight, giving him an awkward slouch. For a moment, I pretended not to see it, because I didn't want to see it. Instead I told him there was a heap of food and that we'd be meeting Skip, Julia, and their boys at the movie soon.

Jason tried to make socially appropriate noises but tripped into a loaded silence.

I sighed. "The inspection?"

"Yeah," he croaked, looking away. While he took a minute before going on, my stomach launched into free fall. Jason explained that when the inspector saw our rustic egg-washing building, he asked the location of our commercial sink, stainless steel counters, plumbed bathrooms, and septic. Our facility was, in essence, a shed on a field with a hose.

Jason had held up his binder, showed him his Xs, the notes, the highlighting. The inspector took it in his hands, squinted at the paperwork, and said, "Oh, I think someone sent you the wrong regs or maybe you're missing some pages, but yeah, this won't work." He handed Jason the binder and began writing out our denial paperwork.

Although we'd technically done everything the state had asked us to do, we still couldn't sell eggs to the public.

"Lu, I've been corresponding with them since last September and not once had any of this been mentioned. They want water testing, a sewer, a hand-washing sink, an equipment-washing sink—even though I have no equipment to wash—and all the plumbing has to be installed by a licensed plumber, with plans submitted to the Department of Labor and Industry," he rambled.

"There's a Department of Industry?" was all I mustered.

"Until all this happens, we can't sell anything," he said with an expression so pained I worried he'd vomit. "I've worked so damn hard for so long and I've kicked down every obstacle given to me but this one. I just don't know."

The reality was that we had rented this particular property based on its ability to meet regulatory requirements, built a building that would satisfy these specific rules, and became stewards of some sixteen hundred living beings with a particular set of constraints in mind. What the state was now requiring was a commercial kitchen with a low-end going rate of forty thousand dollars.

And with the hens starting to drop more eggs every day, we needed it now.

Seeing Jason so low, I somehow became the strong one and moved him to our patio table, setting food in front of him. He pushed it around his plate as I got the children ready to go to the movie, telling them their father was too tired.

"It'll just be us tonight," I said, managing a smile pulled from the well of parental strength.

As I was getting the kids' coats (an essential every month of the year by Lake Superior), I finally gave Jason that beer. I'd lost track of it during the off-kilter moment of the news.

"There are four more in the fridge," I told him. "Play your video game, sleep, we'll talk in the morning."

Though he hadn't made the time in months, Jason liked to play an empire-building game called Civilization. His version is from the late 1990s, slow, and hardly animated at all. When he elects to build something in the game, like a road or a mine, a little pixilated guy will walk onto the screen and shovel for several minutes, while the game catches up. But he doesn't mind the low-tech graphics; Jason's in it for strategy and distraction.

And boy, did he need a distraction.

"Invade ancient Mesopotamia; you'll feel better," I said with a weak smile, my hand on his shoulder.

The rest of the night was one foot in front of the other. The film ran on the screen in front of me, but I only saw the drama of my life playing out inside my head, with no idea if it was a comedy or a tragedy. A thousand-foot freighter, a dot of lights, cruised past on the water, slipping behind the big movie screen on its way to the docks in Superior Bay. It knew where it was going, what it was doing, and that its day would end in safe harbor. That was what I wanted.

When I got home, Jason was asleep, snoring loudly on the couch. When I woke before four the next morning, he was gone.

Chapter 11

Groping with my cell, I located Jason's speed-dial button. It rang a few times as I walked a tight circle from the kitchen to the dining room through the living room and round again.

There was clicking and then, "Hey, Bird."

I heard road noise, suggesting travel at highway speeds.

"Where are you?" I demanded. "It's like"—I looked over my shoulder at the microwave clock—"three forty in the morning."

"Headed out to the farm early to get the minimum chores done, then I have some ideas about this inspection thing I want to check out."

"Ideas?" I haven't even had coffee and couldn't imagine thinking fresh thoughts yet. "What ideas?"

"I'm not ready to talk about it yet." He was annoyingly calm. "But I'll let you know how it goes."

Jason had returned to his unflappable self, which gave me leave to be my utterly flappable self, complete with wild hand gestures.

"Yeah, okay," I sighed. "I can't believe how crazy maddening this whole inspection business has been. You can't give out incomplete rules, have us build a business around them, spending real money, like big money, and then be all 'My bad.' I bet we wouldn't have even rented this particular farm had we—"

Jason cut me off. "This really isn't helpful thinking right now, Lu. I can only look ahead and make the best of the choices in front of me."

I pulled my hand back to stare at the phone. Now I was mad at Jason.

Past eleven p.m., I heard the creaky swing of the porch door. Jason looked tired, but better. I let him start the talking.

"It's been a good day," he said, unlacing his boots. "Been out making friends."

After chores, he'd driven the minivan around the straight roads that hem in the enormous fields in Wrenshall. When he found an ag operation, he'd drive up the long driveway to the farmhouse. These were people, locals, who'd know where unused spaces were.

"How did you even start?" I asked, marveling at Jason's ability to segue into any conversation.

"I'd knock on a door, say I'm the new chicken farmer up on Route 102A, and the government just screwed me," he said, smiling. "It's amazing how fast a mug of coffee appears in your hand with that opener."

It's probably fair to say that most farmers, steeped in a profession predicated on getting things done by one's own hands, have at least some libertarian streak in them. But whatever their political leanings, these conversations, nearly half a dozen altogether, yielded two unused dairy parlors (sterile rooms where cows are milked) near Locally Laid. With floor drains, water, septic, and large basins, these old milking barns might satisfy the egg-washing needs of the State of Minnesota.

I smiled at his genius and thought I might be a little libertarian myself.

The dairy parlors didn't pass inspection.

To his credit, the inspector got out to these neighboring farms and looked at the milking barns within a couple of days. But given their years of disuse, it would have taken too much effort and money to bring them back to current code. But as one inspection was going south, Jason was already on the phone lining up the next.

"I've got another lead," he said, calling me from the minivan as he pulled out of the dairy farmer's driveway, just ahead of the inspector.

"When I took that business class, I met that guy with the store," he said. "He's interested in renting part of his back room."

Jason rarely leaves a room without knowing everyone in it, and that's lucky for Locally Laid. While taking entrepreneur classes at the university, he'd met a man named Shane who, like us, was going into business in his forties. He'd purchased a country grocery store, but unlike us, he'd been working in his field for years and this was a natural progression, a scenario I greatly envied.

But what Jason had recalled from their conversation was Shane's lament that the store was overbuilt for its rural setting and had more space than the business could support.

"Isn't that, like, way far?" When stressed, I hemorrhage words.

Shane's place was in Rice Lake, a solid forty-five minutes from the farm.

"Yeah, it's not ideal," Jason said. "But it's a place willing to take us and we'll just store eggs in the farm's cooler and wash every other day."

That could work. Rod had built a beautiful little refrigerated house on our farm, standing at twelve by fourteen feet. I'd painted it barn red that spring while listening to books on tape for graduate school. I'd marveled at its highly buffed, stainless steel door that we'd bought used from a refrigeration company. That heavy door was likely the nicest thing in the whole operation. I tried to get everyone to call the cooler Egg-arctica, but my family only rolled their eyes at me.

Things moved ahead slowly. It was August before papers were signed and we were able to disconnect the plumbing at our farm facility, nervously transport it down the highway in an open trailer, and bring in the licensed (and very expensive) plumber and electrician to do their thing. Then, before even pushing one salable egg through the egg washer, we had to get the whole deal reinspected. Days quickly added up.

When the inspection problem first began, the oldest flock was just starting to crouch. When someone (preferably a rooster someone) walked past, our rusty brown birds would bend at the knees a bit. That's chicken speak for "Come hither."

Before owning this egg farm, I had only a vague notion of the workings of chicken reproduction. Honestly, I think I was fully an adult before I understood that fertilization wasn't required to produce an egg. Put another way: no rooster needed. (It's a common misconception; if you didn't know, don't beat yourself up.)

Here's a quick run-through on poultry love. While an egg is laid (nearly) daily, it will become a chick only if a rooster's sperm was ALREADY inside the hen before she started forming the egg. So, if a rooster comes into a flock of lady birds on a Saturday and starts right in with, forgive me, the "bow-chicka-bow-bow," a fertilized egg could not be laid until Sunday. Eggs "in process" during copulation will not be fertilized.

Hens have sperm host glands inside the body, which will hold and release a tiny bit of a rooster's reproductive goodness just as the egg is being made, making one encounter's virility extend for nearly two weeks of viable eggs.

While our initial flocks were only hens, we were later gifted a few roosters. They have utility. As hens are busy, head down, foraging, a rooster will watch the sky. At the first sight of a hawk, he'll sound the alarm with his worried crow and send hundreds of chickens sprinting to shelter.

But a rooster will also be there for the romance. Some of the more dramatic males will approach a potential mate flourishing their wings in sultry ways, like a flamenco dancer. Others less inclined to courtship will just wait for an opportune moment when a distracted hen is squatted over a meal and hop on.

To understand what happens next, you need to know that hens and roosters both have only one point of exit, the one-stop shop of orifices—the cloaca, commonly known as the vent. It's the point of waste elimination, laying eggs, and the transfer of sperm. And despite our referring to the male of the species as a cock, they don't have one. (Wait till you trot that nugget out at the next cocktail party.) What they do have is more like a nubbin, called the papilla, located just inside their own vent. Bereft of a functional copulatory organ, there's no penetration in chicken sex, more of an aligning of vents during a piggyback-style mounting, called treading, during which the roosters spray their sperm. For most egg farms with a rooster or two, there's little worry of selling eggs with a forming chick inside the shell. Any fertility in an egg is arrested once it hits the refrigerator, which typically happens within just a few hours of being laid.

All this talk of poultry love is to say that the crouching behavior we were seeing in our hens was a sign of sexual maturity and, more importantly, a precursor to eggs.

That July, the flock was indeed starting to lay. It wasn't the full production we'd been expecting back in late June, but there had been factors. The birds may still have been acclimating to life "on the outside," so to speak. They'd lived through their own *Matrix* movie moment, learning that their entire world, their cage-free warehouse space where they were raised, was not the world at all. That alone could have explained their late laying.

Or it could have been their early brooding, which would set the blame somewhere south at Myron's door. The birds were small for their age, which meant that either they weren't really as old as he claimed or they were developmentally behind due to malnourishment. And let's not forget the wild ride they had to get to our farm. Of course, there was the terrible third option that no poultry professional seemed to be able to confirm but we never could shake: was it something we were doing wrong?

We'd fiddled with the feed ration, set out additional waterers, fussed over spacing in the coop and access to shade—really, anything we thought could make a mature bird happy enough to produce an egg.

Though now, we couldn't legally sell.

Even with only the first flock starting to lay, that was then some 775 birds, and if only 50 percent were laying, that's over 2,700 eggs in a week. Eggs we had no homes for.

Jason started bringing them home at an alarming rate.

"Hey, could you give these a wash?" he said, handing me another of the wire egg collection baskets he'd recently purchased. They would look old-fashioned if not for the blue rubbery coating over the wire.

While Jason leaned on the banister to get upstairs for a nap, I headed downstairs to our mother-in-law apartment. Fortunately, we were between college student renters, giving me a full kitchen to turn into the washatorium. There I sterilized the sink with a spray bottle of diluted bleach water. For the next hour, I'd scrubbed farm life off the eggs—mud, straw, fecal matter, the occasional spots of blood. The small kitchen smelled, well, eggy.

I placed the eggs in the reused cartons I'd been gathering from friends and opened up the refrigerator door to face a wall of stacked eggs.

Legally, these couldn't be sold even at a farmers' market. Plenty of people do, but as we had ambitions to work with grocers and restaurateurs, we were dedicated to following all the rules.

Soon I was shaking Jason out of a loud, openmouthed nap. "Hey," I repeated. "We've got sailing tonight."

He sputtered awake, using his entire palm to rub up and down his face, which appeared rubbery with the unfinished business of rest. He stared, unseeing, into middle space.

"We're a little late, but we can make most of it," I said, talking over his obvious, unspoken preference to stay home. "I've got bread and cheese and beer in the cooler, ready to go."

While we were all tired, I knew rest was not the only thing we needed. All this time spent on the farm was changing us, and I was losing my footing on how to function as a family or even be Jason's partner—not just business partner. We were withering from lack of diversion and shared experience. Plus, we needed to remember why we lived in this brutal, beautiful place—on purpose.

The kids and Jason slumped in the minivan amid an air of mild contempt. I looked in the rearview.

"You remember Windbreakers?" I asked. A notable aspect of our destination, Park Point, was that this seven-mile sandbar jutting into Lake Superior can turn cold without warning. Winds shift over the cold lake and, as my sailor friend Dale says, lips pursed over his open palm, "It's like blowing over an ice cube." A northwest bluster can make a July day suddenly require a wool hat.

Summer is simply different here.

Driving to the sailboat launch, one must pass over the shipping channel via the area's iconic Aerial Lift Bridge, rumored to be the most photographed road structure in the Midwest. Or maybe that's just a rumor Duluth is trying to start. It's an enormous steel girder apparatus, which raises the roadway over the water so ore-carrying freighters, gravity fed their iron cargo from railroad cars on a high trellis, can access the Saint Lawrence Seaway and, ultimately, the Atlantic Ocean. Looking at this whole dance of international commerce—the rail cars, the cargo chutes, the freighters, the tug boats, the lift bridge—it's like a Fisher-Price city operated by hand crank.

Once at the end of the Point, my friend Britt, whose kids go to school with ours, put a glass of wine in my hand and embraced me so tightly I felt more collected than I had in days. Over her shoulder, I saw that Jason was already involved in animated conversation and laughing, Milo was clicking shut a life vest, and Abbie was two-thirds up her favorite tree, a paper plate of lasagna in one hand, a book tucked under her arm.

"How's the chicken biz?" Britt asked.

I dropped my smile and fell into the sad tale of the inspection and before long, a group gathered to hear. Even though I had, over the past few days, grown to a place of acceptance, our friends had to fight their way through the anger, the disbelief, and the urge to fix it.

And it was that fixing urge that became "Outlaw Eggs."

"I'm going to sell them," Britt said. Our mutual friend Maria joined in. True to their word, they took boxes of cartons, getting the buzz started using word of mouth and social media. For nearly two weeks, they ran ovum speakeasies out of their homes.

Later, I would find out that both Britt and Maria were buying dozens and dozens of eggs from us, and mostly giving them away. If I think too much on this kindness, it still makes me teary all these years later.

At the time, I struggled with it. I mean, how do you repay people for that?

Abbie says I should just pay it forward.

A few weeks later, in early August 2012, our egg-washing facility was blessed by the State of Minnesota and Jason dropped off a flat of eggs at the Duluth Grill. That evening I posted a photo of owner Tom Hanson with Abbie holding Locally Laid's very first check. Now with our processing facility set, Jason was free to go out and close deals. He'd pack up our beautiful, brown-speckled eggs and walk in the back doors of grocery stores and restaurants. It's a little ballsy to show up unannounced, but it's also hard not to be at least a little interested in the smiling guy in the dirty ball cap, holding an open basket of dappled eggs.

He'd exude the happy confidence of a man who'd built a better mousetrap and deliver a pitch of how our eggs differed from the ninety-nine-cent ones. It ran the gamut from flavorful yolks and attractive packaging to our environmental work and memorable name. Jason also touted LoLa's "Eat Local" ethos. That *local* in the Locally Laid name was no accident (nor was the sassy double entendre).

No doubt, we'd greatly benefited by the hard work of food activists who'd come before us, adding terms like *locavore* to our common speech. Their efforts were paying off, with retail outlets now seeking out regionally produced food as their customers demanded the freshness, the feel-good, and more variety than commodity producers provide. And all this added up to steady orders for our product.

Certainly our right-name, right-time circumstances might have opened the door for our pasture-raised eggs, but what kept us in restaurants were the cooks. Tom from the Grill later told us that our eggs initially caused a little tension in the kitchen. Line cooks were getting to the egg boxes first and labeling them "FOR LINE ONLY" because our outdoor birds were producing— and I love this phrase—*yolks with superior integrity.*

Cooks weren't tossing out eggs that come out of the shell a bleary mess, ruining their sunny-side-up orders. We'd also heard that bakers liked them for the tall meringue that didn't require cream of tartar, a perking agent for flaccid egg whites. (We giggled around the farm, saying, "LoLa don't need no fluff'ah!"— forgive us, we were tired and punchy.) But most impressive, Tom said that because egg waste was virtually eliminated, our higher price wasn't cutting into the bottom line.

As we started appearing in a few retail stores, I started working the media by sending out press releases and inviting them out to the farm. Getting column inches and TV time was relatively easy given our visual promise of over a thousand chickens in a bucolic backdrop. Add in that all the birds are named LoLa and we've got ourselves a story.

When a news crew was scheduled to come, I prepared Jason with a bunch of talking points—about food miles, our use of non-GMO corn, and that we pledged to plant a tree with every delivery to offset our carbon footprint. And though Jason was a great spokesperson (despite still wearing that grubby ball cap against my wishes), regardless of how hard he steered an interview toward America's broken food system, the story was always pretty much the same.

I can sum it up in four words: OH MY GOD . . . *CHICKENS!*

I've decided Americans are vulnerable adults when it comes to the hilarity of poultry. Apparently, *chicken* is one of the most inherently funny words in the English language. And it must be true; the venerable tastemaker the *New Yorker* said so in a 1948 article. In the piece, haughty H. L. Mencken chastises our "plain people" affection for the humor-inducing *K* sound, making jokes of fine places like Kalamazoo and Hoboken. Ostensibly, this jocularity also mocked the honorable chicken as well.

But for whatever reason, eggs were selling briskly.

In September, we tabled at the Chester Bowl Fall Fest. It's an event that attracts thousands on behalf of the ski program that puts hundreds of children on the slopes every winter.

It takes a special energy to "give good booth," but I was

becoming better at public speaking given that I'd begun wading into teaching as part of my degree program.

While we were selling cartoned eggs from our refrigerated minivan and a few of our sassy Local Chicks Are Better tees, we mostly were there to educate the public on the term *pasture-raised* and get a little brand awareness. There were some unexpected benefits, too.

For one, repeating your story all dang day makes you good at it. Damn good at it. Shivering in the shade and showing the umpteenth person pictures of our chickens eating clover and alfalfa, I was punch-drunk tired and referred to the chickens as "salad-eating poultry athletes," and really, that said it all. It made sense given that when we eat right and exercise, we're healthier. It's the same for a chicken. And a healthier bird would lay a better egg. Universities and USDA-funded studies found that eggs from birds raised this way had less fat (one study found less cholesterol, too) and more of the hard-to-get good stuff like omega-3s and beta carotenes.

This was when I first realized we had learned a lot, like a crazy lot, in a short period of time. People had many questions about chickens and, for the most part, we could answer them. Our painful education on the pasture was starting to take, despite our remedial starting point.

But the best reward for putting oneself out in the big world is the unexpected connections you'll make. Cara, a petite blond woman in her early thirties, came bounding over to us and I immediately saw in her the same exuberant energy Jason possesses.

There was a crescendo of animated conversation involving quick talking and sweeping hand gestures. Cara worked for a good-sized, local distribution company, Upper Lakes Foods, and was hell-bent on growing their line of sustainable offerings. She and Jason set up a time to talk over the winter.

As she and her happy dog jaunted off, I turned to Jason and said, "What the hell are you doing? We don't have the capacity for something like that."

Jason smiled and shrugged. "Not yet."

Chapter 12

It was early December, nearly six months since The Day the Chickens Came, and winter was settling over the Northland. You could feel the change of season in the light; it's thinner with less sustenance. It cast no shadows. And for the first time since I could remember, this magical and exciting pre-Christmas season had felt neither. It was like the sharp intake of fearful breath as the season of death descended.

Ever since we'd talked about starting the egg company, people had asked us how the birds would overwinter. And Jason had confidently and consistently answered that, given a higher-calorie, fatty food called scratch, the girls would be just fine. These were, after all, hardy northern breeds. And when they roosted together, wing to wing, they'd create significant heat, which we'd make the most of with straw bales stacked against their coops. I hoped he was right.

Despite my many hours each week in the egg-washing facility, I still felt like a poser at Locally Laid. Jason, who all but lived at the farm, walked around like he owned the place because, well, he does. And though I technically do, too, it was like the city cousin coming to play in the country. I knew the only way to overcome that feeling was to work through it, literally.

Getting started, I headed off to the shed, our former egg-washing building that was now storage and Jason's nap station.

Jason had asked me to collect eggs while he did some winterizing chores and finished the shelter for an additional eight hundred hens from a farmer who'd sought us out to buy his birds. He was looking to get out of chickens.

At Jason's behest, I donned the too-big kneepads, the kind my father used for laying floors in his home construction business. I tried to exude confidence in them, if only for the audience of me. I was grateful that Jason had already handled the day's more muscle-intensive work of trekking the buckets of feed and the barrels of water. And like many things at the farm, it worked so much better on paper than it was proving to be in real life.

Just as before, watering the hens started with filling the large plastic casks on the back of the four-wheeler from our landlord's hose, then driving them out to the individual waterers and filling them, one by one, from a gravity-fed tube. It was time intensive. And then it hardly worked at all. Despite fiddling with the barrel, the tubing, and the spigot and locating it on the highest possible ground, the water had been barely trickling. Like we couldn't even count on gravity. This added up to hours squandered on this one chore. Jason had been talking about putting in a well at the farm, which is sensible and all—except, you'll remember, we didn't own this land. We were welcome to improve the property as much as we wanted, but we also knew our land-lady would never sell.

Poking around the shed for some eye protection, I saw the *Game of Thrones* book cast atop an overturned bucket, its cover filthy, pages bent. It was clearly a loved volume and I smiled. I'd bought it for Jason despite it not being his typical genre. It was a prescription. The idea was to get him as far away from the prairie as possible.

He'd been jumpy and perhaps more worried about the changing season than he let on. However, I had another theory. It went like this: take a gregarious guy and stick him all alone out on a pasture for fifteen hours a day, seven days a week, with a thousand-plus birds providing a slightly hysteric din, throw in the fear that he'll fail them along with his entire family, and that man will back right up against the edge of reason.

In our prefarm life, Jason read a lot, mostly nonfiction, and I know he would ruminate about these subjects at length. Now he came home too tired and sleep deprived to pick up thinky books, and I was concerned that his excess mental energy was churning up worries where there were none at all. Like that fall, when Jason walked in the door and beelined to me saying, "How's our marriage?"

It was a statement delivered with incredible earnestness. "What?" I asked. "What are you talking about?"

"Like, are we okay?" His eyes wide as he emphasized the word *okay*. "Are you going to leave me, because if I were you, I'd leave me."

His unnerve was contagious. This is the man who loves to

shake things up, and now he sought solace from me, someone who views adventure through splayed fingers.

"Why would you say something like that?" I stepped closer, into the aromatic zone of his work clothing.

"I don't know, I've been out on the pasture, just thinking . . ." He trailed off. His eyes were rimmed red; fatigue and concern hung on him like a horse collar.

Admittedly, I yearned for an existence that made intrinsic sense, at least more sense than this one, but the thought of taking the children away from their father made none at all. And, of course, there was Jason himself. I mean, after all, I am his Bird.

Despite his farm filth, I reached in and folded him down in my arms, pulling his foul-smelling self against my body and rocking him gently. "Jason, I love you . . . and I couldn't leave you even if I wanted to. There's too much debt."

He made a snerty noise and I felt him smile into my shoulder.

"Well, at this rate you'll never be able to leave me," he said, leaning more of his weight into me while I ran my fingers through his wavy hair.

I gave him the book and the edict to read a chapter every day between chores.

"Read aloud to the chickens if you have to, but get totally lost in this world. Stop thinking so much."

And he did.

Having found a pair of oversized gloves, I shuffled outside into the diffuse light. The bulk of the collection baskets were neatly stacked near the farm's feed bin. Despite the cold, this area had the homey smell of grains. I worked apart a pair of baskets and held one up in front of my face. I couldn't help but be drawn to it in the way most people are drawn to simple objects of the past. Perhaps we associate them with a less complicated time with its straightforwardness of task.

These are particularly irresistible when filled with eggs, and LoLa's are various shades of taupe and brown and speckles; the whole montage pulls at something deep inside you. Nostalgia? Romance? I'm not sure, but there's something about the entirely nonhomogenized nature of it that just feels satisfying.

I headed for the first coop, puffing white clouds of breath as I clumped along in my rubber boots. Looking up, I took in our ten acres, framed by the distant barn and abandoned silo. They were dusted with early-winter snow and under the white skies it felt bleak, but oddly beautiful, too.

At the paddock, I made big, exaggerated footfalls, pushing over the bendable fence. I was surprised that my stomach still squeezed as I entered the chicken's enclosure. Despite seeing these birds for months, it happened every time, a fear they would rush me, as they sometimes did when I carried buckets of feed— chickens running atop of chickens in a loud, flustery poultry tsunami.

I hated that I wasn't as comfortable with the working girls as with our home flock. Those backyard birds slid closer to pets than livestock on the critter continuum. And while we were, no doubt, kind to these farm hens, they'd never be as domesticated as the ones in the little coop behind our garage. It was just a numbers game. One can't pick up and coo to some sixteen hundred chickens in the same manner.

At home, the little flock functioned as a classic hen hierarchy with a top-bird-enforcing order. They're all girls, but the female in charge had grown a bigger comb, a testament to her elevated testosterone, bringing to mind a number of "size matters" jokes. I'll leave you to supply your own. But her coop justice was meted out under some human supervision. If a home hen was getting overpecked (in poultry argot that's called *picking* and where the term *picked on* comes from), she'd be pulled out into a separate, back pen of the coop until things settled down. This kind of harassment seemed to be simply part of animal makeup. But on the farm, the only rule of governance was an unmitigated pecking order.

This instinctual code takes the sweetest bird, a chicken who'd seek you out to purr in your lap (yes, hens purr) or snuggle in close to her roost mate, her head slung across her sister's neck, until she noticed some sign of weakness in the chicken next to her. Perhaps a spot of exposed blood or some flag of feebleness that only another bird can perceive. As we learned back when our chickens first arrived, a docile hen can switch up as though receiving a signal from the mother ship and the signal says *kill, kill, kill.* She will obey.

On our website, I'd bragged up the fact that our farm hens function as a real flock, unlike those raised in cages—and they do, maybe more than people realize. Months ago when I wrote that, I was thinking of our home hens gathering up to groom together, giving their feathers a bath, much like cats. In some ways our home hens were properly closer in temperament to house cats than poultry.

At the farm, well, those girls were organized around a set of feral rules, ones that set off terrible clashes. Horrifying sights that made you want to whistle for blue-helmeted peacekeepers as skirmishes turn cannibalistic.

That's why I can be wary upon first entering a coop.

Inside the hoop coops, swinging my baskets, I was relieved at the calm. While there were chickens around, it wasn't quite wall-to-wall birds, given the mild weather. Many had opted to go outside, where the feed and waterers are. Putting these main attractions outside motivates the chickens to get up and move about outside—even if there is no prairie grass to peck at. Exercise is good for a bird, making her less inclined to start a row with her neighbor.

The farther I walked inside the coop, the more aware I was of the farm smell. It wasn't as intense as one might think, considering the sharp ammonia-tinged stink of chicken droppings. It can jump right up your nostrils and bore into your sinus cavity, but because our hens are in these airy coops and go outside all the time, the odor is manageable. The straw helps. So does the breeze. But I'm not going to pretend it smells good, either.

Twenty or so hens were standing on the rounded two-by-four roosts that sat chest high and made up the right-hand side of the building. (Later, we'd convert to sticks, which are more friendly to a hen's feet.) Probably another twenty birds were milling around underneath, kicking in the straw. A dozen plus—more than I'd have liked—were in the thirty nesting boxes, two rows of fifteen on top of each other, drilled into the frame of the coop's left wall. Certainly, enough birds to see the barnyard archetypes at play.

And just as I bent into my egg-gathering crouch at the row of nesting boxes, the Suck-Up arrived. This bird lit on my back, singing her little *coke-coke-coke* song in my ear. I petted her as best I could, given the awkward angle and oversized gloves. She was telling me all the business, while perkily moving her head and shoulders. There's one in every flock, office, and schoolyard.

As I inched my way down the line, mostly on my knees, the tattletale chicken rearranged herself on my shoulder. I was grateful for the ill-fitting kneepads. Despite the strap biting into the back of my knees, it was better than completely sacrificing my canvas pants to the chicken droppings, smearing as I moved from box to box, putting eggs into my basket. Many of them were soiled with fecal matter, despite the straw.

Just to clarify, the notion that eggs are dirty because a chicken defecates as she lays an egg is a falsehood. It's physically impossible, given that the necessary squeezing action of her vagina physically closes off the path from her large intestine. What soils an egg is when chickens lounge in their nesting boxes, treating it like a combination beanbag chair and Depends undergarment.

We overcome this by gathering and changing bedding often and by keeping the girls moving. Modern operations have a clever device that will sweep a bird right out the nesting box in the evening and close the door behind her. This keeps her from messing up where she'll lay her egg the next morning and encourages her instead to sleep on her roost. I see the wisdom in it.

Moving down the coop, I felt crowded. There was probably only two feet of clear walking space between the row of roosts and the row of nesting boxes. If only we could go back to the past and tell our naive, prefarm selves to plan a different poultry shelter. Despite all the thought that went into these picturesque little hoop coops on the prairie, they were difficult to maneuver in and tougher to clean. We really needed a coop interior we could completely dismantle and run a Bobcat-type machine through, rather than the pitchfork and shovel we currently wielded. But how could I complain? If it was hard for me at five feet, four inches, I could only imagine the contortion that six-foot Jason underwent to gather eggs.

Somewhere in the middle of the coop, I came upon another class of chicken, the Untouchables—huddled together in one nesting box. Perhaps they started life smaller, or maybe they were singled out for being a slightly different color, but these skinnier birds were missing tail feathers and most had been bloodied on their stumpy rears. Their combs had been pecked down to reflect submission. Honestly, they were a little hard to look at given the scars of violence they bore, and also the awareness that I wasn't fully a farmer, just a woman with five pampered hens in her yard.

When I gently tossed one of these fragile, lower-caste birds out of the nesting box to expose the eggs, the rest of the microflock hopped out after her. I suspect they felt safer together.

Having trouble cupping up the last two eggs, I decided to take off the enormous glove and grab them. They were still warm.

Sliding over a bit, I came around to a big alpha hen lounging in her own private nesting bin. Fat, with a gorgeous red comb, she puffed up to twice her size when she saw me coming. Taking me in with her beady side eye, she produced a low growl. It was unsettling. I wondered for a moment if she was broody— that's when a bird is in a maternal state and determined to hatch the eggs underneath her. This can happen to chickens that have never even seen a rooster but just have gotten it into their minds that those eggs are fertile anyway.

If her eggs weren't gathered every day, the chicken would sit on a dozen, a full clutch, determined to incubate them for twenty-one days. She'd conduct a sort of sit-in, stop laying, and become vehemently protective of her nest, lined with breast feathers plucked from her chest—like the adage "to feather one's nest." This provides softness but more importantly exposes her skin to better transfer her warmth and moisture directly to the egg's surface.

When a mother hen is on the job, she'll leave her eggs only twice a day to quickly eat, drink, and defecate a giant broody poop, some 500 percent bigger than her normal dollop, that she'll drop far outside her nesting box. Fortunately, the hen in front of me wasn't in a maternal state of mind, probably just warm and comfortable, though I still appreciated the default eye protection my glasses afforded me.

I pulled my leather glove lower on my hand and snuck it underneath her before she could peck. I gave her a little lift, maybe you could call it a goose. There was a loud squawking protest, a rude whooshing of feather tips, and then she landed with a graceless thud and walked away with her beak bouncing forward and a harassed look. It was as if to say, "Geesh, you didn't have to get physical."

Chapter 13

By the end of the afternoon, I'd hoisted eleven baskets each holding about 120 eggs into the back of our minivan, a vehicle tricked out by Jason's cousin Jake, a licensed electrician, and Jake's handy dad, Dan. It now sported holding racks for baskets and a wall to keep the refrigerated air in the back. I reached down and shut the back door with a pleasant clunk, leaning an extra few seconds on the van.

I'm told there was a television show called *Pimp My Ride*, and I think of the Egg Van as a losing contestant. It used to be my vehicle, and I missed it. This was a surprise. While a fancy car would be wasted on me, I yearned for something practical and safe that didn't cause a rubbernecking commotion. It's not like I'd even wanted a minivan, but when Milo was a newborn, Jason and I succumbed to the safety ratings and the large cargo space. My hilarious friend Julee joked that we should have stickers on our vans that read *I'm wearing leather pants in here* to offset the mom-jeans stereotype of driving such a vehicle.

Clearly, it wasn't hipster, but now that Jason had taken my grocery-getter to the farm, it'd been a loss. Not so much the roomy back, though I could cram in the skis, skates, bikes—not to mention the sizable school carpool. But what I pined for, I'm embarrassed to admit, was that minivan's undeniable claim to middle class. Nothing says "I'm a middle-income momma" like a shiny grocery-getter rolling across the parking lot. Over these past few years of declining earnings and uncertainty, it felt like our passkey into the world of "everything's gonna be all right."

Now I was stuck driving Jason's old Honda Accord. You've seen it—I mean, not this exact one, but it seems Japan stamped out countless of these forest-green four-doors in the late 1990s. The make and model was ubiquitous. Jason's particular sedan had just over two hundred thousand miles on it, and I was adding to it every week with my three-hundred-mile round trip to graduate school. It wasn't the car's high mileage that bothered me. It was the dent.

After the last fender-bender he'd gotten into a couple of years back, Jason elected against bodywork, given that we were likely to get a new car in the near future. There was a good-sized crater on the driver's side, made worse by the factory-provided scratch paint over the exposed metal. It wasn't exactly the right shade of green, turning the door into a sad banner of "I tried." I suppose these were all sensible decisions given the age and wear of the vehicle, but there was something downright depressing about driving it that way. Like piloting a rolling flag of failure.

From the farm, Jason drove the Egg Van some thirty-eight miles out of rural Wrenshall north on Interstate 35 through Duluth and back out into the country again, to the store where we washed and packed eggs. While it was a modest addition to the product's carbon footprint in the big scheme of food miles, a number typically in the thousands, it was something we cared about. Our light environmental tread had quickly become part of LoLa's growing value system, but the problem wasn't so much the turn on the odometer as the travel itself. Highway speeds and bumpy side roads broke eggs.

No matter how conservatively we filled baskets or what kind of insulation we'd tucked underneath them, eggs on the bottom got crushed. When I thought of all the hard work that went into creating these brown-speckled beauties, it was hard not to take the loss personally.

I entered the egg-washing facility, a windowless room in the back of the grocery store, with confidence. Whereas Jason was most comfortable out in the field, this was definitely my turf. I asked Ian (Gail and John's son and a member of our paid washing crew—well, they were all paid but me) to fetch yesterday's eggs from the cooler. Jason had delivered those last evening, the final task before his fifteen-hour day ended.

From those two days, we had just over 2,300 eggs to process, putting Locally Laid at about a 72 percent rate of laying from our flock. Our egg counts had been higher in the fall, and there might have even been a week or two in there when the girls surpassed the fabled 86 percent—the target percentage of eggs we'd listed on our business plan.

Later, Jason would say that when it came to business planning, he pulled the rookie mistake of grossly underestimating expenses and overestimating income. Not that we consulted the plan much anymore, given how off-rail our budget had gone with unanticipated entries on the expense side. These included paying farm neighbor kids to put the chickens in their coops on the nights when Jason managed to leave before dark. Those automatic poultry doors never quite worked right. We'd also endlessly—and expensively—fiddled with the chickens' diets, adding minerals and even, at one point, costly fishmeal, hoping for more eggs and less bird savagery—not that we saw sweeping results.

And I was standing in our biggest unexpected monthly expense—our rented processing room—which, along with everything else, fell under the QuickBooks column, "The price of doing business."

I didn't love washing eggs and to be truthful, I still don't. Feeling disinclined to spend four to five hours under the fluorescent lights of the processing room, I had days when I'd get annoyed that we indulged in the entire egg-washing rigamarole. And by we, I mean Americans. Most people don't know that eggs, if left dry, can be stored, unrefrigerated, for several weeks—or so the old-timers say.

And the entirety of the European Union.

Washing an egg scours off its natural cuticle, or bloom, an invisible, protective coating that keeps it airtight and shelf stable—and the reason why EU eggs aren't washed at all. Though, to be fair, the barns in those countries tend to be much smaller than the typical million-hen U.S. operations, giving them a salmonella prevention advantage.

Salmonella is a bacterium that grows in the intestinal tract of critters, or as one book put it, "leads a predominantly host-associated lifestyle." It's one of the most common causes of food poisoning and can be spread when infected fecal matter on a shell contaminates the yolk when the egg is cracked open. This has been the rationale behind washing, and that makes some sense; however, the risk doesn't end there. Salmonella can also make camp in a hen's ovaries, infecting an egg before there's even a shell around it. Hot water can't help that.

That's why government food safety websites recommend that we thoroughly cook our eggs, killing all germs. But not all eggs are consumed that way. Even if you personally avoid dipping toast tips into a soft-boiled breakfast egg or eschew classic Caesar salad, think of the last time you made chocolate chip cookies. Did you taste-test the dough?

To prevent transference this way, the United Kingdom requires that their laying hens get vaccinated against salmonella. And even though there's no law requiring this in the United States, we do it. It's less than a dime per chicken and helps me sleep at night. After starting their vaccination program in 2009, the Brits dropped their infection rate to only 1 percent in their flocks.

In 2010, when the USDA updated its regulations and placed its faith in washing and refrigeration, some 550 million American eggs were recalled. A nationwide salmonella outbreak occurred just after the new regs rolled out.

There are other disadvantages to relying on washing. For one, it runs through untold gallons of water and then, once bereft of their bloom, eggs require refrigeration from washroom to cold truck to chilled commercial display cases and finally, your home. Some reefer trucks, as they're called in the biz, have two engines—one to fire the pistons and one to crank the cold, adding to a product's diesel impact. The whole thing feels wasteful.

While I knew my way around the washroom, I tried to avoid "running" the crew myself, a group of four typically made up of two teens and a couple of adults. The crew leader responsibilities include both quality control of eggs, which I'll explain more in a minute, and fulfilling the orders of flats—square, open cartons holding thirty eggs and boxed for wholesale customers like the Duluth Grill, two local bakeries, and a hospital, among our earliest commercial clients. Then there were cases of our typical cartons, sporting Jason's nontraditional slogan of *Get Locally Laid*, with fifteen dozen to a boxed case, headed to our local co-op and one nearby grocery store.

I'm relatively content placing dirty eggs into the front end of the egg-washing machine, the AquaMagic V I mentioned earlier. When I first saw this narrow, eight-foot-long invention, standing at chest height on metal legs, I was reminded of *The Wizard of Oz* and the scene of the Tin Woodman getting scrubbed and polished by large circular brushes, a device clearly inspired by egg-washer technology.

Here's how it works: Dirty eggs are put on an intake ramp in a repetitive manner, one by one by one. As each is placed, the ramp doubles as a candling device, forcing light from below to illuminate shell cracks or blood spots. Cracked ones are plucked off right away, as any break in the shell is a potential inlet for bacteria. Sometimes I'd see our young helpers hesitant to throw away an otherwise good egg with the smallest of cracks.

"You gotta think of the little kids and old folks who'll eat these. We're watching out for them," I coached.

The rejected egg made an unpleasant *spleck!* sound as it broke in the trash bucket.

Eggs with blood spots also need to be tossed, but honestly, we hadn't seen any. These red flecks that artistically contrast with a yellow yolk have nothing to do with fertilization—though it's a really good guess. It's actually a spot of hen's blood, an indication of a burst blood vessel when the yolk was forming inside her body. It's safe to eat, but Americans tend to be put off by evidence of where their food comes from. I get it.

After the candling inspection, the egg is picked up by a plastic chain conveyor and gently tumbled inside the machine, where it's hit with jets of hot water and pushed between two scouring brushes. That this water is significantly warmer than the egg is crucial because of the porous nature of shells. Cool water would create an internal vacuum, pulling surface bacteria into the egg, contaminating it.

Progressing down the line, there are more hot-water jets with gauges ensuring a steady ninety-degree temp, ending with a splash of light chlorine water (fifty parts per million) to disinfect the shell, per state mandate. Then a final blast of air from a fan and the egg tumbles out of the machine onto a wide, white tray.

The person on the exit end, the crew leader, must then pick up and examine each one and determine one of four fates. If it's still soiled, it takes a return ride through the AquaMagic; if a hairline crack is discovered upon second inspection, it's tossed; if it's somewhat lumpy and unattractive, it gets put into the Ugly Box, destined for our house. We eat uglies. A perfect beauty of a brown, speckled egg wins a ride on the weigher.

Eggs pop out of the washer one every second, leaving little time for rumination.

Turns out, I'm not fast enough to evaluate all these factors simultaneously. Not even close. I've had moments resembling the classic *I Love Lucy* skit where Lucy and Ethel can't keep up with their factory line wrapping chocolate confections. The pair comically stuff sweets into their mouths and hats and down their brassieres to hide their slow progress.

These are all poor options for eggs.

But that doesn't mean I haven't gotten flustered. I've been known to rewash perfectly clean eggs or lose my train of thought completely and drop them on the tile floor. The soft *thwap* can be sickening, like denting a small skull. Jason thinks of it more monetarily, like he's throwing a quarter (a generous approximation of what we gross on each egg) into the garbage.

Clearly, I wasn't built for the hot seat.

Eggs deemed clean are placed in a plastic egg container

and walked over to the vintage Egomatic, our midcentury egg weigher. Eggs aren't sized by volume but rather by mass. The metal machine has a series of brass counterweights that, once hand-calibrated, will pass an egg to a metal hand until it finds the appropriate chute for its weight and then drops it down to be packed in as a Jumbo, Extra Large, Large, Medium, Small, or Peewee. I want to state right now that while Peewee sounds like a term I'd create, it's actually the technical classification by the USDA for any eggs collectively weighing less than eighteen ounces per carton.

Farmers would call these pullets, small eggs from newbie egg layers, and though there's a bit of a foodie cult fascination with them because of their concentrated eggy taste, there's not really a market for them, mostly because of the cost of cartons. Not that an individual carton costs so much—actually it's about fifteen cents—but rather that you must buy them in bulk and pay for them up front. That can be a five- or six-thousand-dollar outlay, not to mention the designer's time and our time getting it state approved. Then you'd have to beg for shelf space in the grocer's dairy case, because honestly, when was the last time you bought medium eggs? Or even saw smalls or peewees in the store? Certainly not a lot of pullet eggs in the display case either.

The truth is most Americans do not buy anything smaller than a large, partially because recipes call for that size. I've wondered if that's one of those looping problems. Americans don't buy mediums, so American cooks don't create recipes for them.

Jason has solved the smaller-egg problem by selling them to restaurants for less, but not the pullet-sized ones. In addition to uglies, we also bring home tiny eggs and the rare ones so large the jumbo carton won't close. I hate seeing those because I worry about the bird that laid it. It had to hurt.

There would be more of those bigger-than-jumbos as the months passed, since older birds lay larger eggs. Back then, our birds were youngish, so we collected mostly mediums and larges. As the hens matured, we saw larger eggs. In addition to being less desirable to consumers, jumbos are fragile. Incredibly so. A chicken produces the same amount of shell covering no matter the egg size, so you can think of it as blowing up a balloon. The shell just gets thinner, more delicate, and prone to breakage.

Packing the last cartons, I looked around at our four-member crew and oscillated between marvel and horror at how many hands these eggs were passing through. You could say it started with me that morning as I plucked them out of nest boxes and then brought them here, where they were placed into the washer, inspected, often rerun and reinspected, then put into the weigher and packed. That could be counted as nine separate touches. But the handling really started before, if you think of the moving of fences and daily recharging of feeders and waterers.

It's enough to make an efficiency consultant drink bourbon straight out of my work boot. And it's one of the primary reasons that eggs produced this way are some three times as expensive as typical eggs. Sure, our feed is more costly and, unlike confinement operations, we actually need farmland—both real costs. But I can tell you the bulk of that money goes into people, paying them.

Conventional eggs, the $1.20-per-dozens from confinement-caged operations, are often not touched at all. Like never. Once that chicken is positioned in her cage, she consumes her feed from a conveyor in front of her, depositing her egg contribution onto another belt behind her. These eggs are then motored via belt over to an egg-washing facility, where they are automatically loaded into the washer, inspected by machine, and packed a dozen at a time by an articulated arm using a suction-type technology, releasing the eggs gently into their carton with a hydraulic swoosh. These types of operations don't have people working as much as supervising.

In comparison, we're positively medieval.

After rolling our last cases into the cooler, ending a solid four-hour evening with just over two thousand clean eggs, I felt thoroughly damp and fragrant with an eggy cling. I wanted to go home, but it was time to power-wash the machine. Whatever bits the pressurized water ushered to the floor—pieces of straw, chicken droppings, an odd feather—it all had be mopped up and surfaces disinfected. The evening ended with the unglamorous tossing of the broken-egg trash bag into the Dumpster behind the store. It was hard to keep track of how many we threw out every washing session. It would fill a good-sized Hefty bag halfway. It felt like a lot.

On the cold drive home, uncomfortable in my moistened jeans, I caught myself thinking, "This is a little like working a factory job," and then realizing it was *exactly* like working a factory job. I tried hard not to count how many years of liberal arts education Jason and I have between us, instead concentrating on the hot shower waiting for me at home, followed by a jelly jar of wine. Maybe two.

The next morning, after getting the children off to school and after Jason was long gone to the farm, I set to family errands, starting at the grocery store. While pushing the cart and hamhandedly tearing coupons from the supermarket circular, I was prioritizing my day. Milo had occupational therapy in a couple of hours, and while the contract copywriting for the ad agency had mostly dried up, I still had a couple of freelance gigs to get out the door. And there was always plenty of reading and writing to do for graduate school.

At the checkout, I watched my items flow down the conveyor belt and was coming up with my bagging strategy—what to put in the cloth ones I'd brought from home, which meats to slip into plastic bags, when the cashier announced (a little too loudly, I thought) that my card had been declined.

"Declined?" I said back, like I was suddenly a foreigner and unfamiliar with the term. In a way, I was. There have been times when we've forgone trips and home improvements because of finances, but there had always been money for basics. And there's nothing more basic than food. Plus, I had always been keenly grateful for it. I'd long performed a post-grocery-shopping ritual. Walking to the then Toyota minivan, now busted-up Honda, I'd put a hand to my heart, offering gratitude, a shout-out of thanks to the universe for allowing me to feed my family. And now as this woman handed me my useless card back, I wanted to say, "But there's some mistake; you see, I've been *appreciative*."

Doing quick accounting in my head, I tried to discern what had happened. Then I remembered: health insurance. We were back to buying our own. That bill was nearly the size of our house payment, and I was guessing that through the magic of electronic funds transfer (and my poor planning), it had scooped the remaining money out of our account.

"Oh," I said.

The cashier was not unkind, but I was nervous anyway. Quickly working through my disorganized wallet, I saw the company credit card—the one I'm not to use for personal expenses, ever. Jason had been resolute and clear on this.

I handed her the card. It worked.

As I walked to the car, my breathing was off-kilter, but I forced myself to slow down and somehow be grateful anyway. The Lake Superior beach stone I wear around my neck, an artistic gift from Gail, was clutched in my hand. It served as both a warm comfort object and talisman.

Hours later, long after I'd fed the kids, I stood by the kitchen window watching for the egg van to pull into the alley. When its lights illuminated the garage, I ran outside to meet Jason. It was cold, with some snow on the ground, and while I preferred to talk inside, I didn't want the children to overhear. The garage's motion-sensing light popped on and we met in the spotlight, where I blurted out my checkout story.

"What's going on? What's wrong with the farm?" I asked. "I keep going back to our initial projections." I said this knowing all the factors that knocked those numbers off track months ago.

Jason could barely hold a facial expression through his weariness, his clothes smeared with filth, and me, at this moment, asking for his comfort.

"I don't know," he said quietly. "It's . . . so much harder than I'd thought."

Though I'd long suspected, this was the first time he'd said it aloud. I imagine he felt duped by the seemingly simple nature of the pasture system. I flashed back to his "Itsy Bitsy Spider" hand gestures at the Mexican restaurant, representing the beautiful cycle of sun, the grass, and field rotation. The imagery did not align with the dirty, nearly broken man in front of me. He felt betrayed by the sustainable models based on warmer climes that ran, at least partially, on intern labor.

"But nothing really bad is going to happen. I mean, we're not going to lose the house?" Our house in the Cities had finally sold a few months back, and now that we were down to one home, I was keen on keeping it. But I'd really only asked to hear him refute my doubts. I mean, we were having a tough moment, but we were still of middle-class stock; things would work out. They always did, right?

"I don't think so." Jason repositioned, swallowing. "I guess, I'm not sure."

Standing in the cold, I saw the world through a Jell-O fog. The goose bumps that had sprung up on my arms and legs seemed to be pulling my whole essence upward and out of my body. If I'd had eaten dinner with the children, I would have surely vomited.

"I'm sorry, Lu. It's just so hard."

That night in bed, I couldn't shut my mind off. I tumbled the realities of the business like pouring sand from cup to cup, trying to get our Big Gulp–sized expenses not to spill over the sides of our Dixie cup income. That was when I saw it. On the wall was a spider, about the span of my palm, with a full and round abdomen the size of a fifty-cent piece.

Spiders make me light-headed in a way other insects don't— the way their robotic scuttle is always faster than anticipated, and how at once they're angular and pointy and bulbous, promising an explosive, wet splat if met with a bluntly wielded phone book.

The spider skittered across the wall and launched itself onto the bed, legs extended. The sharp breath I'd gulped exhaled as a forceful, unbroken scream.

As I rocketed now to a standing position on the bed, six eyes took me in—though not arachnid ones, but the anxious human eyes belonging to my family. They were filled with concern for me, but also, no doubt, for themselves.

Because night terrors were back.

They'd last visited me just as the chickens arrived, back when I was having trouble eating, an unusual condition for me. Typically stress is a call for ice cream and hot, buttered cinnamon-sugar toast served with tea and cream. But this wasn't that kind of tension.

It sat in my gut, a palpable mass of anxieties leaving little room for food and displacing bitter stomach acids that seemed to enjoy the ride up my esophagus, given the frequency of my acid reflux. I knew I was losing weight because both my pants and my peers told me so.

But there was nothing to do, other than to ride through our rocky startup.

Chapter 14

January swept in, bringing Canadian air masses and frigid temps with it.

"It's stupid cold, and my phone's going to die," Jason shouted over the percolating bird chatter. "The wellhead froze again this morning."

After bantering about the foolishness of digging a well on rented land, we'd decided that a few thousand to save Jason's back and solve a myriad of farm headaches was worth it.

Until it wasn't.

"How can the well be frozen? Isn't it heated?" I was incredulous.

After we had spent the money to punch (that was the preferred verb of our well digger—*to punch*) an artesian well and then protect it with an insulated fish house and space heaters, how could the hoses have the audacity to freeze?

Last fall, when the drilling company was finally ready to attempt a borehole after weeks of scheduling conflicts, we'd rocked on the balls of our feet, hoping for water. It was anticipatory agony. We were paying for this hole in the ground whether anything flowed through it or not. The truck's giant piston had been cranked into the sky, standing erect like a fire truck ladder, and the auger bit, the width of a salad plate, was set in place. They were willing to drill up to two hundred feet and if there wasn't water then, well, that's just tough luck and thank you for the check.

After two hours, they ground through the eighty-foot mark and finally, mercifully, hit flow. Water came gushing forth from the ground and we were happy as the Clampetts, packing the truck for Beverly Hills.

But that was weeks ago, before winter bit in.

"Bird, you can't really complain about it freezing. The thermometer here says minus thirty-three."

"Oh."

This sank in. Not only was Jason working outside in these inhumane temps, but the squawking members of our investment portfolio were currently huddled together with only tarps and hay keeping their tiny hearts beating against the cold, never mind keeping them laying eggs. It was like never-ending winter camping on the prairie.

And as one might expect, egg production dropped. It fell well below the modest dip we'd anticipated, and what eggs we did find were often frozen in nests, despite us collecting every few hours.

We were working our winter plan, feeding our northern gals scratch, keeping their coops buttoned up against winds. They'd be fine. It was what we did with our backyard hens with solid success and what every piece of sustainable ag literature preached. Plus, we lit the hoop houses. When the sunlight shortens in the winter, chickens get less action from their pineal gland, which releases the hormone that triggers egg production. It's all part of their circadian clock. On the farm, we use commercial poultry LED lights, but a simple twenty-five-watt bulb on for some seventeen hours a day is enough to keep birds laying through the season.

Prior to Locally Laid, I understood the dark season as much as any office worker could, having run from one climate-controlled building to the next, until I bundled up for a few hours on the ski hill or trudged through the woods on snowshoes. But nothing could have prepared us for our first winter outdoors with our fifteen hundredish chickens. Coincidentally, the 2012–2013 winter would go down as one of the coldest, with strings of days below zero, and the snowiest on record with a series of late-season snowstorms. It felt like a season actively trying to kill you. The last major accumulation was on April 22, a full seven months after our first flakes in September.

And it wasn't just cold and snowy. It was disheartening. That time of year, the sun is sluggish, and it scarcely backlit the low, white sky that covered the Northland. It felt like an over-turned Tupperware bowl had descended over the entire Twin Ports region, ensuring that no excess light or joy leaked in. But the mild, seasonal depression I suffered was nothing compared to the farm challenges.

Although the well was a huge improvement over our old water drum method, we still didn't have underground piping to the coops. This meant using a hose, pressurized by the well's electric pump, to fill five-gallon buckets, and then towing those buckets on a trailer pulled by the Kubota to the coops.

But when the longer hose from the wellhead would freeze, Jason had to fill each bucket directly from the flexible piece of three-inch-diameter plastic piping through which the well was channeled. This was an onerous task that stretched out for hours in the cold. The shortness of the plastic pipe meant he had to take the water containers off the trailer to fill, then lift the forty-three-pound buckets up and back onto the trailer. This action was repeated over fifty times a day, meaning Jason was literally moving a ton of weight.

He gave up on using lids for the water containers because the thin plastic would simply shatter in the extreme cold. That meant that as the water was pulled along in the trailer, it would slosh. As it sloshed, ice would form on its floor, which, trip after trip, increased the weight. And the heavier the trailer, the more likely it'd get stuck.

Freeing it often took a half hour or more of shovel and sledge work, hauling snow and whacking ice. As winter wore on, Jason stopped using our standard trailer, opting for a smaller one, really more of a cart. It carried fewer buckets and required more trips but was lighter and less likely to get mired. Later, when even that small trailer would become too weighed down with ice, Jason could haul only a couple of buckets at a time in the small back area of the Kubota.

Once he finally arrived at the hoop coops, Jason would pick up the heavy buckets, one in each hand, and carry them with careful steps over the flexible fencing—avoiding the slick patches—and into the chicken paddocks to fill the waterers. Despite his care, Jason did fall a few times and ended up back at the chiropractor. It was Chicken Antics on Ice, Poultry Capades on the Prairie.

The other difficult task was keeping the water from freezing. We'd purchased warmers and changed out our plastic waterers for old-fashioned steel poultry founts that could sit on the electric heating discs. Picture an overturned metal bucket on a big saucer. There's a handle on the flat top of the inverted bucket that, when lifted, reveals a water-holding metal tube inside. These two sections, the tube and its bucket sheath, can often freeze together as water creeps up the metal sleeve and freezes there. To get them apart, Jason clanged them vigorously with a hammer, a tool often covered in frost itself.

There was no winning at the farm that winter. It was a race to keep the birds alive, calling on Jason to dig in with Shackleton-like tenacity, just as he did all those cold hockey mornings. Even the county extension agent, a government-paid farmer resource, shook his head during a courtesy call and said, "No one should have to work this hard."

During this time, unbelievably still in early winter, Jason would come home hours after dark, vacant inside his body. He was still losing weight and wore exhaustion like a rubber mask, making him prone to lulls of silence when eating dinner. Conversational banter was beyond him.

At one meal, Jason managed to say, without looking up, "We do all this work to have them on pasture and the chickens won't even go outside into the snow—even on the warmer days."

On days above freezing, it'd be good for a chicken to get some exercise, stretch a bit, and relieve some of the stress of being "all cooped up." Otherwise, cabin fever would inevitably circle back to the bird aggression we'd been trying to tamp down for months. Jason explained that the hens would come to the edge of the coop and gaze out, as wistfully as a chicken can muster.

"Well," I ventured, "can chickens see snow?"

I had a theory that chickens, with those reptilian eyes and jungle brains, simply couldn't register that white swath the way we do. My guess was that for hens, it's perceived as the end of their world and no good would come from stepping off. The next weekend I was at the farm, I laid some straw on the pasture, putting an especially generous heap by the door, where they'd be making that giant first step. And surprisingly, it worked. Something during that ugly winter started going right.

In other attempts to do what I could, I started cooking like a 1950s farm wife. Jason, out in the January cold, was wasting away before me and clearly needed a heavier diet. I went online to learn how to prepare pork loins or brisket or whatever was on sale in the meat department.

Jason came home one gray winter afternoon and looked more forlorn than his standard-issue level of downcast.

"What's up?" I asked, seeing him hunched in the entryway.

"Oh, LuBird, I . . . I lost my wedding ring," he said. "Must have slipped off my finger on the pasture."

I could not suppress my groan. I was angry at LoLa, irked that she'd managed to work Jason into such a thin little nub that he couldn't maintain a ring on his finger. Also, I fought against reading too much into it, seeing it as a clear sign that the farm was nullifying my marriage. In all likelihood, it was gone forever. A bird probably saw a shiny on the prairie and ate it.

While part of Jason's weight loss was the incredible physicality of it all, it also came from the stress of disappointing customers. The eggs had gained a certain popularity. The Duluth Grill wanted everything we could get them, while cartons were also in demand at the handful of retail outlets selling Locally Laid. The birds were using everything they had to simply stay alive.

Though I knew the girls weren't in full production, it was during a Wednesday night egg washing that I truly understood how little they were laying. It was just Jason and me. We'd sent our helper home early and it struck me that the task was getting done in no time at all. I looked at the skimpy number of baskets on the metal cart.

"Jason," I shouted over the whir of the AquaMagic, "how many eggs were we getting a day last fall?"

"I don't know," he said, not looking up from his task of sorting eggs on the other side of the machine. "Sometimes fourteen hundred?"

As we washed roughly every other day, that meant over twenty dirty baskets filled the cleaning floor.

That night I could count the two days' worth of egg baskets on one hand.

I did some quick math and took a sharp breath. According to my multiplication, we were bringing in about a hundred dollars a day—before paying any of our expenses. Feed, rent, workers, electrical, insurance—the list went on.

"Jason, we must be losing thousands every month."

"Yeah," he said, still not looking up.

"Jesus, Jay, you're killing yourself and we're hemorrhaging our life savings," I said, panic now elevating my voice. "Honestly, there's got to be an easier way to make NO MONEY."

This business was going under.

We were going to fail big in front of my family, his family, and our community, which had rallied around us and the farm. I thought of the volunteers building in the spring rain. We would disappoint our children, whom we'd gotten so excited about this venture. How would they ever trust us again? How would we trust ourselves?

"Laying just went to hell, Lu. I don't know what to do," he said. His tone was flat.

"There's got to be something to do. I mean, everyone is making money on this but us." I gestured around this rented washroom and mentioned our feed supplier.

"What, then? You want to quit?" Jason now looked at me, square on.

And there it was.

I then understood that Jason could not pull the plug on this without me explicitly telling him to do it. He'd dragged me into this, and without my express permission or perhaps even a direct order, he couldn't stop trying.

"I don't know," I said quietly, looking down at the dirty egg in my gloved hand and back to the man barely standing beside me.

For sure, a part of me wanted to end this nightmare and put us back into the comfort of office working. A big part. But there was a surprising something else going on in my head and chest. What if we were, as they say, three feet from gold? I thought of all the hurdles we had overcome. What if we just had a few more to go until we'd figured it out?

I opened my mouth, not entirely sure what I was going to say.

That was when our friend Lee Ann walked in. The food buyer for a Duluth hospital, she gave Locally Laid its first official institutional order. Now we'd hired her son to help wash eggs a few hours a week. She'd popped in to say hello before driving him home.

"Hey, how's it going?"

She was bright, happy, clean—states I was certain I'd never be again.

Our faces said it all. Lee Ann, with her easy smile and affirming presence, managed to cajole us out of our funk. I don't recall what she said exactly, but I know that by the time we'd sprayed out the machine, mopped the floor, and bleached the last surface, we were feeling better. At least good enough for us not to finish that earlier conversation.

It could be that Lee Ann inadvertently saved Locally Laid.

Because shortly after that, right when I was sure Jason could not go on another day, something happened.

If this were a movie, the camera would pan across the bleak prairie to Jason, dirty and broken-spirited, struggling to heft buckets over the electric fence. He might have just slipped on the ice, spilling that hard-won water as he dropped the full container on the snow. But something would catch his attention. He'd be looking squinty-eyed at something in the distance. Something . . .

puzzling. It would be hard to make out against the dazzling backlight of the winter's low sun, reflected off the snow. But a figure would step nearer and nearer until Jason would hear an iconic greeting he'd recognize anywhere.

"Dude . . . what's up?"

Brian had arrived to work the farm.

Chapter 15

While we'd been busy starting Locally Laid, life had had twists and turns for Brian, too. Mimi and Jason finally got all the paperwork straight for Brian's longtime girlfriend, Soon, to come to the United States. Within weeks, they were married in a charming ceremony in a Minneapolis park. She wore an intricate beaded gown, a dress she saw in a wedding magazine and sewed herself—in a weekend.

The couple was living at Mimi's townhouse as Soon started the hard work of acquiring a third language and Brian endured seasonal jobs. He hadn't landed a full-time position since returning home from Asia a year and a half ago.

Just as Jason's job loss had once freed him up to attend to Brian in Cambodia, Brian was similarly available to help his brother through his winter terror. He started coming up to Duluth a couple of days at a time, sleeping downstairs in our trundle bed. Then two days morphed to four, and, not long after, he was spending the workweek with us, while Soon and their young baby, Mya, were south in Minneapolis.

The brothers working together to water the birds transformed the task, while still hellish, into a doable misery, something they could limp through together. Jason had no doubt that if Brian hadn't come along, neither his back nor wherewithal would have made it through the winter.

Walking into the house one evening, I ran into Brian, looking dirty and tired in his muddy winter gear. I immediately felt guilty that while I'd spent my day as a teaching assistant in a climate-controlled room, he'd been out doing physical labor in the cold.

"Oh, Brian," I said. "I'm sorry, it's so damn hard out there."

"Aw, you know, we're all learnin'," he replied in his laid-back-dude tone. "Experience. You just can't Google that."

It was a pearl of prairie wisdom. I stared, lost in this wise observation about life and process, turning this modern-day adage around in my head. Then, just to prove he was still Brian, he quickly followed it up with a joke involving Willie Nelson and oral sex (which is something you actually can Google).

Brian started streamlining the operation, finding ways to make the chores more efficient, getting the farm on a tighter schedule for everything from pasture rotations to feeding to coop cleanings.

"He's just a better farmer than I am," Jason would say about Brian. "He has more field sense than I do and, more importantly, he freed me up to work on the business."

When Jason says "on the business," it's a distinction from

working "at the farm," just spinning the daily plates of necessary labor some fourteen hours a day. With Brian there, Jason had more hours and more energy to course-correct the venture as a whole and look at creative ways to make enough money to survive that first year.

Certainly, without Brian, we couldn't have looked at Locally Laid beyond the day-to-day struggle.

It was that January when Jason tossed a brochure on the dining room table.

"I think I should go to this," Jason said.

"The International Poultry Conference," I read aloud with some skepticism, saving my most mocking tone for the tagline, "Meat Me in Atlanta."

I pushed up my glasses and brought the document closer to read the fine print.

"Honey, everything about this sounds like Big Ag to me."

The folks who attend this event count their birds by the hundreds of thousands, if not millions.

"Good," Jason replied, "then they must know something about chickens."

Setting the flyer back down, the date caught my eye: late January—in Georgia.

"Hey," I said, "I think I should go, too." It had, after all, been a long winter.

A few weeks, one plane, a train, and two buses later, we were the non-business-attired minority in Atlanta. With my red canvas Dickies jacket and matching Doc Martens knockoffs (shoes Abbie had lent me), we were a colorful standout in a dark sea of twenty-five thousand blue and black suits—and a few dozen Amish.

We strung lanyards over our necks and hit the exhibition floor, where I felt like a gaping rube at the 1829 World's Fair. The booths were massive, with machinery, movie screens, and large, rotating overhead signs. Some exhibitors even trolled for conventioneers with women in tight skirts who'd gladly bend over to polish your shoes . . . with their hands. Some things you can't make up.

"I should go talk to those exploited young women about union organizing," Jason said with a wry smile.

My look said, "Yeah, keep walking, farmer."

There was plenty to see. Colossal, hissing machines the size of our living room rolled trays of eggs to big metal arms injecting them with who knows what at the rate of thirty shells a second—then conveyed them away. This was robotic technology right out of the *Terminator* movies.

Compared to our hands-on operation, where the biggest piece of farm equipment was a Toyota Sienna, it was like we'd landed on some technically advanced planet. One getting bigger every day. Even so, Jason remained convinced that the big business of chickens had things to teach us about automating some of our systems, while we kept birds on pasture.

He went from booth to booth explaining our business model, and while the salespeople were polite, they didn't have much to offer the outliers from northern Minnesota. I could see them struggle with what to say to Jason as I liberally ate the candy off their tables.

It was evident we were in the small-farm minority in Atlanta, but I knew there were a few others—namely, the Amish crowd. They were the only group that stood out more than we did, and I was dying to talk to them. Despite the women's bonnets and simple prairie-style dresses and the men's robust beards, we probably had more common ground than not. I sidled up to a group, but when I opened up my mouth I couldn't think of anything to say that wouldn't sound like, "Hey, that little hat of yours tells me we'd have a great conversation about natural methods of poultry production."

There were an awkward few seconds. I sputtered a false start, smiled, and then let them go. I'd wager they were relieved. I know I was.

Feeling pretty done, I was trying to persuade Jason to bail with me into the Georgia sunshine. Then we turned a corner in the enormous convention center and came upon the European companies. They weren't attending the conference to sell their pasture farm products, but they eagerly dug out their iPhones to show us movies of grass-based operations in Germany.

Jason's face became animated watching the small screen. It was birds, outside, yes, but also with feed running on a conveyor belt, giving each hen a fair shot at dinner; integrated nipple watering systems, which would eliminate dirty and frozen waterers; and nesting boxes that gravity-fed eggs onto a covered conveyor directly to a packing facility. This made for a much cleaner product. And as most EU countries do not allow egg washing, there's a big incentive to farmers to keep shells unsoiled.

This stirred Jason up and got his mind whirring on a new

vision, which he talked me through as he bought a seven-dollar slice of pizza on the convention floor—though, honestly, he was so lost in his monologue, I don't think my presence was actually required.

"I'd want a barn with several huge doors on all sides to keep birds rotating on fresh paddocks," he explained between unattractively huge pizza bites. "And get good regrowth on resting fields. You sure you don't want some?" He thrust the half-eaten slice in front of my face.

The irony of Jason talking sustainable production while chewing a meal made of processed commodities was simply too ironic not to tease him about.

"Hey, I have limited time here," he said. "I'm not going to drive around town looking for local food."

He was right. Big Ag is easy to support without ever meaning to. It's cheap, convenient, well marketed, and direct-dialed into our brain stem. (I say this as a gal who likes her bag of salt and vinegar potato chips as much as the next person.) And because of cheap diesel, it's everywhere, too common to even notice, like the air we breathe.

On the surface, cheaper sounds better. Feed the world, right? And given that populations are only increasing, I used to think that while sustainable agriculture is a nice idea, we probably do need all the sprays and mammoth machines of the industrialized system to really keep people fed.

Yeah, that's not true.

There's a heartening side-by-side study between conventional and organic crops conducted over thirty years by the Rodale

Institute. It found that organic yields not only matched conventional yields, it outperformed them in years of drought—using 45 percent less energy while building healthier, water-holding soil.

And there's evidence that the industrial model isn't even meeting its world-feeding claim. According to the United Nations' Food and Agricultural Organization, one billion people are currently starving and an estimated additional three billion are classified as "not well fed" in our world right now. Although we've long produced enough calories to feed the world, much of it isn't actually feeding people. Renowned food journalist Mark Bittman writes that a full 5 percent of crops become biofuel, a third goes to feed meat animals, and another third is just plain wasted as it works its way up the rungs of the food chain.

Every step from source to processing to restaurants and households incurs some loss, but this is especially true in American households, where food is cheap and treated as such. One report cites that we've more than doubled our food waste from 10 percent of total garbage in 1980 to now over a fifth of all our refuse.

That's something we could change by altering our lifestyles to redirect calories where they need to go. However, there's another catch with conventional farming that's not an easy fix. Big Ag takes bigger water. And that's a resource we finally have to pay attention to.

While I'd understood that conventional farming relied heavily on nonrenewable oil to make chemical fertilizers, power machines, and transport crops, I didn't get how their farming practices made for thirsty soil. Loaded with industrial saline from chemical fertilizers, conventionally farmed fields do not absorb and hold water nearly as well as organically farmed land. It takes an estimated four gallons of water to make one gallon of ethanol—until you count all the irrigation that went to the corn crop used to produce it. Then you're talking a drastic seventeen hundred gallons of water for that one gallon of fuel. It certainly takes the feel-good out of the pump. Of course, livestock takes more than its fair share of water, too. Nor are chickens exempt from this hefty water footprint.

We're starting to get it.

In 2015, California governor Jerry Brown instituted the mandatory order for cities to cut back on water consumption by a hefty 25 percent. While surely this will be helpful, it seems small when one takes into account that most water in the Golden State, a huge 80 percent of it, is used for crops.

Because fuel is cheap, it's been a real deal to grow fruits and vegetables in California's reliably hot climate, rather than risk less advantageous regions, even ones far closer to home. That's why your broccoli likely travels several states to your grocery's produce department, rather than from the farmer a few hundred miles away. Likely, he or she could grow it, but not as cheaply or dependably as their Sunshine State counterparts.

It's called production concentration, and it happens all over the United States. Take Iowa and Minnesota. These two states produce some 38 percent of all the pork in this country as tiny Delaware strains under a two-hundred-million-plus broiler bird industry. That works out to more than eight hundred chickens per square mile. This massing of production not only strains water supplies but also creates a feedlot waste problem, which falls to government to figure out. And, of course, it makes it difficult for local producers all over the country to compete with mass production's vertical integration and huge economies of scale.

I finally dragged Jason onto an outdoor patio near the convention center where we enjoyed a beer in the seventy-degree weather—a hundred degrees away from frozen water and stuck trailers. It was a much-needed respite from being, as Jason said, "Mother Nature's bitch."

"Now that my brother's at the farm, we can really move forward," Jason said, palming the beer glass. "When we get back, we'll start trademarking everything so we can go license this brand."

Jason's always about a dozen steps ahead of me, and of himself too.

"Yeah, people are just clamoring to start their own LoLa farms and break frozen chicken shit off waterers, too, Jason," I said between sips. "In fact, they'll PAY us to do it."

"No, I don't mean like how we do it now, I mean like at our new farm."

"New farm?"

Jason explained that now that he'd seen the semiautomated way the German and Dutch keep hens, he could clearly envision a better farm. Not just better, but entirely different.

Having based Locally Laid off idealistic pasture-method

books and videos, Jason felt duped. From his sore-back perspective, sustainable agriculture's nostalgic celebration of hard labor wasn't just regressive, it was literally hurting those doing it. To do just the bare minimum of farm chores—lifting buckets of water and feed—at our small operation requires twenty-six hundred pounds of lifting. That's a ton and then some. Every day.

"I want to take the best of the commercial chicken industry—like the barns and watering systems that won't freeze—and marry that with the most important part of what we do—getting birds foraging and exercising on pasture," Jason said while ticking off these items with his fingers. "It wouldn't just be better for me; frankly, Lu, it'd be better for the chickens."

Jason was talking fast, clearly forming big ideas about a cutting-edge retro operation. Now that Brian was sharing the workload, Jason had regained the energy to pick up nonfiction again. His bookstand held volumes about radical economics, historic shifts in markets, and green industry. That was how he decided that Locally Laid, an operation not exactly on the edge of ruin but one with a hell of a good view of it, was going to be a "market disruptor."

This kind of change is defined as so radical that it helps create an entirely new market, like one for eggs laid by pasture-raised birds. The new method can then displace the dominant methods of producing eggs from caged or barn-bound hens who eat nothing but grains.

The car industry provides a good example of this. In the 1970s, during the gas crisis, American auto manufacturers were still churning out huge, fuel-guzzling vehicles. That was when the Japanese swooped in with their high-mileage cars. It's fair to say, we've been playing catch-up ever since.

Jason wanted to be the Japan of this scenario. He had a vision of our small, substandard, clay-splattered farm in northern Minnesota rocking the entire overconsolidated and calcified egg industry in America.

"LuBird," he said, "this is going to be good; I can feel it."

As he enthused over changing agriculture in our country, I was still stuck at "new farm."

On the bus rides back to our far-off hotel, I got to thinking that with no immediate business or kid worries, we were kind of on vacation, although a poultry-themed one.

When we got back to our room, I entertained flirty thoughts, and while Jason prattled on about German-automated pasture systems, I dug in my suitcase and disappeared into the bathroom. I emerged wearing boy short underwear and a newish push-up bra—a miraculous garment with integral hoists and pulleys that cheated gravity while making the most of my minimally stretch-marked, middle-aged stomach. Months on the all-stress diet plan did have some advantages.

Jason scribbled in the notebook while I tried to get his attention, my hands on his waist.

"Put that down," I whispered. "Pay attention to *this* Bird."

"I'm thinking if we get rid of those inefficient feed troughs, we could cut our corn bills by maybe as much as a third," Jason babbled on. "And the chickens wouldn't have to compete for food . . ."

I took the notebook out of his hands.

"Oh," he said, returning to the present. "You look . . . wow."

I engaged him in a kiss, and just as I was guiding him toward the bed, he broke off and said, "Oh, Lu, I'm sorry. I . . . I've just got too many chicken-y things on my mind, you know?"

I closed my eyes and threw myself backward onto the king-sized mattress with a sigh.

Mistress LoLa had gotten to him first.

ACT 4

Pecked

VERB: struck with a beak; to be beak slapped

Chapter 16

That early spring, I was walking the halls of yet another convention, the natural-fiber yin to Atlanta's meat-pun yang—the Midwest organic conference named MOSES. While we'd been the eco-hippie radicals in Georgia, this Wisconsin gathering was a celebration of sustainable agriculture. And we were clearly the moderates in the room.

The morning started with affirming yoga followed by a big breakfast buffet of steel-cut oatmeal, real maple syrup, and organic fruit. And unlike any industry gathering I'd ever attended, it was kid-friendly. Thanks to a grant from the Farmer Veteran Coalition (an organization that would later help us buy a used tractor), we piled the whole family into the Accord and headed to La Crosse.

While walking Abbie and Milo to the kid zone for a day of shirt dyeing and bird feeder building, we took it all in. The exhibition hall was in a cement-bleacher stadium, circa 1980. Booths were set up along the basketball floor, visually engaging with snapshots of crops and livestock, but without the jaw-dropping one-upmanship of the Big Chicken conference. No rotating electric signs the size of a swimming pool above our heads, no tight-skirted women eager to shine our work boots.

The conference standouts were the young fervent types, a faction of counterculture eco-growers who give me hope for the world. Certainly, MOSES accounts for a higher concentration of cotton garments, dreadlocks, and warm personas than I'd ever experienced in one location. Here my red Dickies coat went completely unnoticed in this plaid and beard-heavy constituency, likely the original assembly of the modern lumbersexual.

At a coffee kiosk, a pale and tired-looking knot of twenty-somethings huddled around its fair-trade beverages. I pegged them as likely participants in the conference's unofficial nightlife. These lively gatherings, usually involving live music and always at a bar, could be found by listening in on the right conversation or its modern equivalent, following Twitter hashtags. It was rumored to be an energetic singles scene, with farmers of both genders on the hunt. Of course, this all went down as we snored openmouthed in the hotel room, unable to break out of our early-morning wake-up habit.

Clutching hands, the kids and I walked single file through the smiling crowds, nodding and excusing ourselves as we swam through the cement corridors on a current of affable people. Attendance numbers followed the national trends, citing a surge in small producers. It's a bit counterintuitive when the attention and tax breaks come to monstrous agribusinesses, but small operations are actually a huge growth sector in food production.

They make up more than half of the numbers of farms in the United States, grossing anywhere from a thousand dollars a year to a few tens of thousands. Often they sell their produce using a direct-to-consumer model: think farmers' market or a food share from community-supported agriculture, better known as a CSA. According to the 2012 USDA census, there are nearly 24 percent more farms following this model since 2002. And sales from these operations spiked from $812 million to a dizzying $1.3 billion in a decade. That's an increase of 60 percent in income. Small, it seems, is exploding.

That's some needed good news among data that also tells us that the big farms aren't just large, they're massive. The top 10 percent of megafarms suck up some 70 percent of the cropland. The *Washington Post* found that the mightiest 2.2 percent of these vast operations control a full third of the nation's available acreage. That's like taking a mud pie representing the farmable earth in this country. Cut it into thirds and take out one honking slab of it for under 3 percent of farmers. It's a crazy visual.

Certainly Locally Laid doesn't fall into this "mass class" category. In fact, back at Atlanta's International Poultry Conference, our having added to our flock to now have just over a couple thousand chickens made us idealistic bumpkins. But at MOSES, people were blown away by our farm's enormity, to the point that I avoided talking numbers at all. A few also eyed us with suspicion when we talked about semimechanizing our next farm to mitigate the brutality of the daily labor, as though we were cheating.

After kissing Abbie and Milo good-bye, I slipped in late to a breakout session, finding a place to lean along the back wall. The speaker, a confident woman with an impressive list of academic credentials, was querying the sixty or so attendees about health insurance. "Who in the room carries a policy?" she wanted to know as she paced. About half of the attendees, maybe a few more, raised their hands. When she asked those covered by a spouse's off-farm job to lower hands, only a couple remained high.

"Off-farm income" was once seen as a financial Band-Aid during the Great Depression and other economic downturns. But now it's the norm. According to USDA research, outside work accounted for 87 percent of American farmers' median income. But even the big operations with gross sales over a quarter of a million dollars were relying on off-farm money for 25 percent of their agricultural household income. Having this regular outside money—not to mention fringe bennies and retirement packages—isn't just nice, it's foundational.

While the speaker clicked through emotive photographs of farmers with their verdant produce, she narrated the stories of illness or injury that drove them from their fields. Sometimes charities supported them through tough times, but often there was nothing. And that's the rub. Despite the good news that more farmers are serving their localities, mostly selling direct to consumer, with superbly grown and tasty real food, it's not clear if they themselves are sustainable. It's like feeding people well has now become a pleasant avocation and seemingly as lucrative as a yard sale. And with about as much security.

Meeting up with Jason for coffee between sessions, we chatted about which ones we'd hit next.

"I'm signing up for the free twenty minutes with an attorney," he said, running his finger down the program. He was eager to learn more about trademarking and licensing, a concept he hadn't forgotten since our last conference. "What about you, Bird?"

I was distracted. "You know, I'm loving it here. It's absolutely wonderful and hopeful and . . ."

Jason waited, wide-eyed and attentive, a hand grazing my shoulder.

"But these conferences make me feel like we don't fit in anywhere. It's like, because we're a little bigger and not interested in selling at farmers' markets, we're the odd farm out."

Just a few weeks after MOSES, I fell into some off-farm income myself as marketing director of the Glensheen mansion, a grand historic home and tourist attraction on Lake Superior. Sadly, the thirty-nine-room museum is better known for its tragic murders and reported hauntings than its impressive early twentieth-century household collection.

But there was an unexpected benefit to working there. Because the mansion is part of the state's university system, I found myself crossing paths with many academics. This included Randy Hanson, a soft-spoken and well-liked professor in his fifties who runs the university's sustainable agriculture farm in Duluth. One afternoon, helping his students unload pumpkins for Glensheen's autumn event, I relayed to him my out-of-place conference experiences.

"Lucie, Locally Laid *is* between small and big ag," Randy said, handing me a gnarly, misshapen gourd. "You're a midlevel producer and that's a whole topic of academia—agriculture of the middle." This magical new term encompasses farms grossing between $100,000 and $250,000. Not big, not small.

I could have dropped my gourd.

It makes sense that the term *middle agriculture* hadn't caught on. There's nothing dramatic about the midpoint, not when there's the pocket-sized vigor of farmer David to root for or the sweeping might of big-machine Goliath. Honestly, there's little that's sexy about either middle age or Middle Ag. And lucky me, I'm both.

While walking over the bridge that led to a twisty path to my carriage house office, Randy gave me some book titles. Over the next few weeks I read long scholarly articles, which led to a domino effect of new reading material. I attacked them, pencil in hand. I'd like to tell you I was writing keen observations in the margins, but mostly I was underlining the many words I needed to look up and rewriting sentences from academia to lay English.

And so began what I call the Education of Lucie B.

I quickly learned that Locally Laid had settled into that not-so-sweet spot, the linty navel of agriculture's center. Mid-sized farms, like awkward teens, don't fit in nicely anywhere. They tend to be too large to sell all they produce directly to the public but lack the scale to acquire the big-boy toys needed for large-scale commodity production in the global marketplace. Farms like us, middlers, have to sell to other businesses, be that directly to grocery stores, restaurants, or a distributor.

It's not easy.

And there are fewer of us every day. Between 1997 and 2012 the number of these not-too-big, not-too-small types of operations declined by 18 percent. That's over 130,000 farms that have shuttered the barn doors, cued the tumbleweeds.

Economy of scale comes into play as big agribusiness wields its buying power to make purchases in great bulk. It's like they get to shop at the Sam's Club of farm needs while places like Locally Laid run to the corner bodega, lucky if anything is in stock much less at a good price. Certainly huge corporations buying a hundred thousand pullet birds get a better cost per chicken than our order of hundreds. Large enterprises, counterintuitively, also have lower labor expenses made possible by investments in large machinery. These costs are spread out over their enormous production, meaning that a half-million-dollar contraption doesn't seem so expensive when the farm complex has a million laying hens, cranking out just under three hundred eggs per bird a year.

Locally Laid experiences its scale disadvantages every other afternoon, when we pay four workers to wash a few thousand eggs—a feat accomplished by industry's big players in minutes with minimal human involvement, other than the flipping of a switch.

So why does this consolidation in farming matter?

It matters for a lot of reasons, but especially to the ten million Americans living in rural poverty. That's nearly one-quarter of the nation's impoverished. And I mean right now, not in some sepia-toned, yesteryear memory.

When midsized operations go away, it doesn't just affect one family, it dings all parts of the regional ag industry, like grain mills, feed stores, processing plants, and farm jobs. So there's just a lot less money floating around a community. This erodes tax bases, affecting schools, roads, and livability issues. As the Agriculture of the Middle project puts it, the loss of midsized farms "threatens to hollow out many regions of rural America."

There's another important consideration. When land is farmed by a family that intends to pass it along for generations, you'd like to think it's treated like a prized possession and not some agricultural sweatshop, churning out maximum production, season after season. Think of how people treat a rental car versus one they own forever.

As I read in bed one night, with a snoring Jason beside me, a major hurdle became clear. Midsized farms struggle to establish important business-to-business relationships with restaurants and stores. It's not that these growers can't supply fantastic product; it's about getting that fantastic product in front of consumers, be it cafe diners or grocery cart pushers. To get there, they have to go through kitchen and dairy managers who are generally used to filling one order form with one big food distribution company, then writing one check.

Convenience is the enemy.

Tom Hanson, owner of the Duluth Grill and an early adopter of sourcing regional ingredients, explained that for all the feel-good, taste-good upsides of buying direct from local farmers, there's a trade-off. It's plumb cumbersome. It means giving up the well-practiced elegance of uniform boxes rolled out of a big semi and neatly stacked in the cooler on a tight weekly schedule. Instead, one gets a series of loosely planned drops of baskets and hampers and recycled crates filled with harvest-fresh produce—sometimes washed and processed, sometimes not.

If you've ever worked in food service, you'll understand that a restaurant can, at times, fall into a magical rhythm. Each worker serves as a symbiotic member of a meal-generating organism. It pulses with life, there's a flow, it's a dance. And when the pace hits that exquisite pitch where one can almost see the tendrils of energy making their way through a well-run kitchen, to be pulled out to deal with the farmer at the back door is as welcomed as interrupting good sex.

Setting down the book in my lap, I felt a motivational shift, an inkling of what my role in farming actually could be. Beyond egg washer and getting LoLa to show a little media chicken leg, I felt a swing from wanting to sell these eggs to keep my family afloat to a desire to kick open an entire market—not just for our product but for the whole middle-agriculture sector. It was a Peter Parker moment, like I'd just learned that I was some kind of agricultural Avenger, minus the superpowers or Lycra.

Switching off the light, I spooned against Jason and smiled into his back. We weren't going to change this industry with our fledgling agricultural abilities. No, this was where those seemingly useless liberal arts skills, along with Jason's hockey chutzpah, would come into play. And though I wasn't sure yet exactly what I was going to do or how I was going to do it, it was all percolating in the dark.

"I just want to jam the term *middle agriculture* into the lexicon—can you imagine it?" I told Jason, now awake. "We could

use our farm to start a bigger conversation about midlevel producers and their econo—"

"You said it!" he shouted, now abruptly upright. "That's the
first time."

"Well, I just learned the term," I said.

"No, not that." He smiled. "You said, *our* farm. You always
refer it as 'the farm' or 'Jason's farm' or 'that damn farm'—you've
never called it *our* farm before."

I smiled sheepishly. Although I wanted to deny it, he was
likely right.

Chapter 17

It was during that long, warm fall that Jason called me from the road. Through an acquaintance of an acquaintance, he'd been asked to visit a group of farmers looking to morph their in-barn operation into one resembling LoLa. We'd hoped they'd be interested in licensing our brand—agreeing to raise eggs to our standards while renting our well-publicized name and professional look to break into markets. This would provide us with some much-needed income, while saving them thousands of dollars and lots of years building a brand of their own.

I envisioned Jason meeting with these skilled folks, coaching them on pasture rotation and best practices against predators. I could show them how to score news stories and optimize photos for use in social media. And later, we could tutor them in how to work with dairy buyers to create room on their local store shelves.

"Bird," Jason said, sounding high-pitched on his cell, "their farm is exactly what I want ours to be, with modern feeding and watering systems and the ability to get the girls out on pasture."

While I couldn't exactly picture the setup, it sure sounded better than Brian and Jason hauling heavy buckets around the farm. Locally Laid wasn't just low tech; it was Schlep Tech.

But while the farmers wanted to work with us, they had something other than licensing in mind—contract egg production.

"Contract production?" I said, shouldering the phone to my ear while simultaneously grating cheese and monitoring Milo's homework progress.

"They'd produce the way we do; we pay them right away with cash for their eggs and we get them out into the world, in front of customers. We already know how to work with retailers and distributors around their region," he said.

"Oh," I said, worry in my tone. I circled a problem on Milo's worksheet and gently tapped it with my finger. Milo shrugged lethargic shoulders, prompting me to push over a math-motivating M&M.

"They should really just rent our name. It would be so much better for them."

"Lu, they want nothing to do with selling or marketing. Contract production was their idea."

It hadn't occurred to me until right then that farmers, real salt-of-the-earth humble producers of food, aren't likely as forward-leaning into the world as we were. That they might not be interested in shilling out a press release with humbling follow-up phone calls, word-raking through a social media post for twenty minutes until it's crisp as shaken linen, or simply ghostwriting for a chicken at all. It's not easy navigating that sweet spot between witty engagement and out-and-out edification. This kind of work isn't for everyone.

"But isn't contract production like modern-day serfdom?"

I'd run into this term in conjunction with broiler production, the meat bird industry exposed in food documentaries, and later through reading research from the Pew Trusts. This mass production of meat chickens has been hailed as the most complete example of vertical integration in agriculture. A parent company, acting as an "integrator," say a Tyson or a Perdue, will have a stake in or outright own the hatcheries that brood the chicks, the industrial farms that raise the grains, the mills that blend their feed, and the processing plants that later take mature chickens from birds to meat to nuggets. Then they're all packaged and loaded on their big rigs to be trucked across the nation.

Now this kind of assimilation makes for a very efficient, bounce-a-quarter-off-it type of system.

By 1955, nearly 90 percent of the broiler industry was run on contract production, and at the time it seemed a decent venture. There were a number of processors, allowing farmers to shop around for the best contract handshake deals back then.

It's changed considerably. Now broiler contract production seems about as advisable as a payday loan, given the expensive technology needed to be competitive and the fact that four big companies—Tyson, Pilgrim's, Perdue, and Koch—dominate the market share.

And the way the modern system is now structured, the industry really has the chicken farmer by the short feathers.

Let's do a quick walk-through, hitting only the highlights. Handshake deals are now history, replaced with ironclad contracts. The grower must build a barn to the integrators' specifications. Fine. Say a farm family invests some three hundred thousand dollars in a building with a loan over fifteen years. And while many growers sign long-term contracts, few get any guarantee of how many flocks they'll get annually. The birds are owned by the integrator (as are the feed and antibiotics), and these chickens are in and out of their barn in under fifty days. The only thing the farmer has legal possession of is the loan debt and the chicken shit. And I'm not even making that up.

Contractually, growers are left to figure out what to do with tons of manure, and since they're not growing any grain for the hens under their care (low-price feed is trucked in by the integrator), they have little agricultural use for it. Many growers don't have fields, only enough land to build the warehouse. And with multiple growers located near the processing plant, there's more waste than the region can handle. Improper management of this broiler litter has led to polluted waterways and federal cleanups.

Everything, even the tourney system that pits growers against each other to determine how much they're paid, is slanted against the grower, but that isn't even the worst part. Big Chicken, at any time, can demand expensive upgrades to the facilities, which the farmer pays for—or the integrator might withhold birds. If there's complaining, well, a grower might wait a very long time between flocks or get terminated altogether. And what of their barn payment then? There's more than one documented case where farmers have broken the cycle of servitude by suicide.

"Jason, you need to encourage them to license, steer them away from contract production," I barked into the phone. At this point dinner prep had come to a standstill. "For one thing, they'll be less beholden to us, plus they won't miss out on all the fun marketing parts."

"Well, Amish farmers might have a different definition of fun, Lu."

"Amish?" I said, bewildered. "But I thought you said their farm is modern."

While we were still dragging fences across uneven fields and gathering eggs on our knees, an Amish farmer who freely rejects the modernity of zippers had us grossly outmechanized.

Well, don't that sting.

And at that moment, just as Locally Laid considered the responsibilities of contract production, another proposal was on the table.

I'd been out biking with Milo when his chain fell off. While attempting to wrestle the greasy thing back on, a couple approached. The gentleman, athletic and a bit older than me, helped me feed the chain back in place. He held out his now-dirty hand and introduced himself as Philip. After some friendly chitchat, in which he learned I was "the Locally Laid lady," he indicated he was a businessman in industrial manufacturing and an investor. It just so happened he was looking to add a new business to his portfolio.

Jason was eventually brought in and we started a conversation that went on for a few weeks, just as we were moving forward with a partnership with the Amish farmers. After learning our business model of sourcing and selling locally, Philip concluded that we should avoid contract production and build a much bigger farm here in the Northland. He envisioned scaling up to three huge barns with eight thousand chickens each—and a generous salary for Jason and me.

"This doesn't feel like Middle Ag," I said to Jason as he tied his work boots, getting ready to leave for the farm. "How can we sell that many eggs within our region?"

"What he's proposing, Bird, is really more of a factory model," said Jason. "I'm not even sure with a flock that size we could actually get all the birds outside."

He rubbed his face and sighed. "But I don't want to live through another winter like last year, either."

We tussled with the offer, wondering if there was a way to accommodate a bird's natural instincts to roost and dust-bathe and forage outdoors in these kind of numbers. And I'll admit it: the salary, and the financial relief it would bring, tempted us.

But just as we'd come to the difficult decision not to accept the offer, Philip called and abruptly withdrew it. No hard feelings. Obviously, it was for the best, but still it was like rejection from a lover—even one you were about to dump.

This left the partner farm option. And after rejecting a scale-based growth model, contract production seemed like a better way for Locally Laid to bring enough money to keep our farm going, satisfy the demand for our pasture-raised product, and grow beyond the reach of our northern farm.

That is, if I could become comfortable with the model.

Jason and I engaged in a conversational dance around the

ethics of contract production. I would bring up some fresh area of freak-out and Jason would calmly lay out why we would not morph into immoral money fiends, harnessing fellow farmers as minions.

But in my ear I could still hear the respected keynote speaker at MOSES warning about small, sustainable operations like ours getting too big, too commercial, too corporate.

"Isn't that exactly what we're doing, Jason? Are we going to be like some evil egg-pire stamping LoLas across the country-side?"

He sighed a little. I'd probably asked some variation of this question at least a dozen times recently. He spoke slowly and all but donned a sock puppet in his effort to convey his message.

"LuBird, no. You're confusing us with McDonald's. We're not like that because those kinds of franchises truck their Mc-Food product all around the country. This Amish family will be buying feed *in their area*, encouraging a crop farmer to plant more non-GMO corn—*in their area*. They'll buy farm implements from an *area dealer* and they're going to be making more money than they did keeping their birds inside for cage-free eggs—and where are they going to spend that money, Lu?"

"In their area?" I sheepishly offered.

"And," he said to close his argument, "we're the ones taking all the risk. We still have to find locations for these eggs."

The truth of it squeezed my stomach. This contract meant Jason would have to sell eggs in the big Minneapolis and St. Paul metro area, a place where we'd had only a little media play or industry connections. Sure, we were in one co-op there now, but suddenly there were a lot more pasture-raised eggs to unload.

Jason put his arms around me. He was too kind to say it, but we both know that I am not a visionary.

"This way we can meet egg demand and get farmers farming," he said. "We're doing the right thing for Locally Laid and everyone involved. It's all coming together, Bird . . ."

I smiled wryly.

"This could be an opportunity to redefine contract production," Jason went on. "You know, make it not suck."

I decided that given that nine out of ten farm households in the United States require some infusion of off-farm cash, this sourcing and selling of good eggs might as well be ours. After all, Jason had spent months figuring out how to create a market for our different kind of product and the complex transportation system to get our eggs into shoppers' hands. Why shouldn't he have a side job doing that work for others?

I sighed, leaning into him. "I'm always just asking, 'What would LoLa do?' I don't want to let her down. She's, you know, like a friend of mine."

It's true, despite how utterly diagnosable that sounds. Through LoLa, my sassy and sustainable logo hen, I was able to write more tenaciously about eating local, animal welfare, and the treatment of the planet than I ever could have without her. More than anything, I could feel, building through her, a desire to be the spokeschicken for Middle Ag. And while contract production seemed difficult and scary, it also sounded like agriculture of the middle.

We signed a contract.

Before those birds came into lay, Jason set about finding homes for these future eggs. We got help in the strangest of places. My friend Eric had been wearing our farm's Local Chicks Are Better tee when an acquaintance took a liking to it. It was through a friend of that new shirt admirer that Jason was now sitting in a room with the dairy-buying executive team of one of the Twin Cities area's biggest grocery store chains.

This was not the type of operation that had a back door accessible to winsome farmers. Jason had been trying to get in touch with these buyers for weeks, but not until a peer contacted a decision maker directly was Locally Laid granted an audience.

It felt that reverent.

Just mentioning the meeting with this grocery corporation made friends swoon. They made us feel like our financial struggles would soon be behind us.

Jason knew this wasn't going to be easy. The feeling in the room, as he described it, was "combative and extractive."

"What do you mean by that?" I'd asked him when he came home, dazed.

"They made it really clear that they're the Big Dance and just talking to us was an enormous favor."

In return for a little shelf space, the chain of stores wanted concessions. Big ones. They were insistent that our prices be higher than we'd prefer. After some go-rounds, it became clear this price hike was not to benefit Locally Laid, but to take us out of competition with their house brand.

But, they'd said, raising our price would also help offset the mandatory discount. What I didn't know is that when an item goes on sale, the store retains its full margin on the product (in this case, 25 percent) while the producer gets flogged by the sale deduction. Also, we were expected to roll out paid ads to support our brand—and these ads were to mention that our eggs were available at the particular chain.

Of all the policies, what was most difficult was their remittance system. Unlike most of our vendors, this big corp was only going to pay after thirty days. To a small operation like ours, cash is critical. It's all about the flow. We had myriad expenses and were often left playing it close to the edge, driving checks directly from the mailbox to the bank. And we didn't much like the thought of scratching to keep our farm and feed partners paid while this industry behemoth made interest on our money (little as that was).

After a couple of hours, Jason emerged from the room feeling both educated and beat up, but with a handshake deal to start with a few stores. He followed up the next day with an e-mail and then, a week later, a second. They never responded. Nor did they return calls.

Sometimes business just goes that way.

And though it took weeks of traveling back and forth to the metro, Jason found homes in smaller stores for all those future cartons. Of course, things would fluctuate and there'd be sweaty weeks when we wrote checks to our partner farms for eggs we couldn't sell (sending them off to food shelves). But despite a few hiccups, it was slowly starting to work.

We were officially farmer integrators, on a quest to redefine the term for the better.

Chapter 18

Jason carried disappointment like feed buckets, his shoulders stooped under its weight. After the bulk of the morning chores were done, he'd left Brian at the farm. He had an appointment in the glassy downtown offices of the university's Economic Development Center. Wearing his biohazard of a frayed ball cap, Jason arrived to talk through another rewrite of our loan application. There'd been no bites to fund the new farm, a place where we could build or retrofit a structure for chickens to enjoy semimodern feeders and waterers that wouldn't freeze, while also having big doors with pasture access. Loan officers, it seemed, were leery of planting modern money into our cutting-edge retro agriculture.

And though we were deep into the lush heat of summer, Jason lived in the shadow of winter coming. The prospect of another cold season, exposed on the prairie, terrorized us. Neither Jason's nor Brian's body could take it.

But as he'd been getting up to leave the meeting, our university-provided business coach, Curt, indicated he had one more thing. Turning his computer monitor to face Jason, he said, "You might do well with this."

It was a small business competition put on by Intuit, the makers of our accounting software. And the prize was a stand-out: a commercial aired during the Super Bowl. While winning wasn't likely, Curt pointed out that entrants received a free trial of the online accounting software. This was a motivator.

I'd remembered seeing the contest announcement floating around Facebook and had meant to enter. But now that I was working a nearly full-time job and busily typing out my graduate thesis, I hadn't followed through.

In reality, I'd forgotten about it.

I looked it over the next day during my regular Locally Laid office hours. These began well before the sun or even the chickens were up, but I needed this time to work and write before heading to Glensheen. Scanning the rules, I saw that all I had to provide was a short paragraph about our business story and a photo.

"I can do that," I said to myself, and after writing and rewriting the blurb down to the short little nub the character count allowed, I threw it on the contest site with our logo. It couldn't hurt.

"Maybe it'll get us more exposure and that'll help our loan prospects," I called out from my indented writing spot on the couch where I worked my laptop. Jason crouched over the dining room table and made a disparaging noise. "We needed that farm loan to happen yesterday," he said, eyes never leaving the invoice he was writing. Seasonal worry hung over him like a cartoon snowstorm.

A few weeks later, we received a mass e-mail saying that we'd made the cut to the next round, us and fifteen thousand other businesses. Now we needed four essays and an introduction video to continue on in this four-round voting contest. We were now competing for clicks.

"We've been talking about making a video forever," I'd said.

"Yeah, okay." Jason was distracted by his own writing project, a livestock grant. "See if Beau wants to do it."

I'd already started work on a quick script for my former ad agency colleague. He was now making a go of it as a freelance videographer, and though he was headed to a conference in Minneapolis, Beau graciously agreed to stop at the farm on his way out of town. He and Jason shot the piece in just a couple of hours, which Beau edited overnight, in lieu of sleeping. And having recently taught himself banjo, he set the farm footage over his own upbeat soundtrack.

The next day, with my laptop illuminating the morning dark, I smiled at the screen, then laughed and immediately hit play again. The farm was in its summer glory and Jason earnestly performed my script, even introducing himself as Locally Laid's Head Clucker. I'd thought for sure he'd balk at that. Somehow the quick video captured all the reasons why we chose to work so damn hard while still being fun and cheerful. At one point, as Jason voiced that our outside birds were "poultry athletes," Beau brought a sprinting chicken into a hilarious slow-mo. It both squeezed my heart and delighted my funny bone. And Jason spoke with such sincerity in the clip, I even forgave him for wearing the damn hat.

As the contest required online votes, Matt, our designer, got busy creating whimsical, downloadable posters with the tagline *This Chick Is a Game-Changing Fowl* and stickers to slap on every egg carton going out the egg-packing room door. I started mentioning the contest daily on social media and leveraged the buzz into a few TV and newspaper stories, always good for business and a great opportunity to introduce the concept of midlevel producers and our strange world of Middle Ag.

"Lu," Jason said a few days later, "type the contest site into Google. What do you get?"

Intuit never divulged vote counts, but my search engine results showed our Locally Laid contest page indexing as the third-most-viewed contestant on their website.

"That can't be right," I said. "We've tainted the results by visiting the page too often. I'll check it out later."

After a Saturday morning of errands, I kicked my windshield wipers on high against the rainy fall weather and drove to Duluth's Canal Park. It's one of the best viewing spots for freighters on Lake Superior, which makes it the perfect location for a bank of hotels. Pulling into a parking lot, I ran through the storm to hover at the lobby's guest computer. Water dripped into my shoes as I smiled apologetically to the woman currently using the terminal.

When the PC opened up, I slid into the chair and punched the contest name into the search engine. Here, on this completely unbiased machine, Google showed Locally Laid as the second business listed on the site.

"Holy shit!" I said louder than I intended, causing a flutter of newspapers and the uncrossing and recrossing of legs across the lobby. A kick of excitement flared in my chest. Did that mean we were one of the most-viewed links on the site? People were voting for us? This couldn't be right. There were thousands of other small businesses competing, many established for years. We'd sold our first egg only thirteen months ago. But this search showed us directly behind NORML, a huge organization over forty years strong, striving to legalize marijuana. They're huge. One hundred eighty-six thousand people follow their Twitter page, while we were barely breaking two hundred.

Clumsily, I fished my phone out of a pocket and snapped a grainy shot of the screen. Texting it to Jason, I wrote, *LoLa's almost as popular as pot!*

That was when we really started feeling a buzz.

Soon I couldn't go buy a gallon of milk without three people stopping me to tell me they were voting LoLa. I'd stand there uncomfortably as I secretly stashed some other egg brand under giant packages of toilet paper in my shopping cart—Locally Laid eggs were constantly out of stock, and Jason never wanted to bring any home from the farm that could be sold. And since I wasn't in the washing room much as I worked full-time off farm, the uglies went home with the crew.

It wasn't just me. Strangers would see Brian wearing his LoLa tee at Target and ask if he'd clicked that day. And our blog, previously read only by the chicken-obsessed and Jason's mother, was now getting views by way of Texas and California and Florida as hundreds and, later, thousands of people followed the contest link to our website.

From the comments people were dropping in digital space, it

didn't feel like this was about just one small farm in northern Minnesota. I had the sense they were voting daily for a whole new food system or at least curious about alternatives to the one we have. Perhaps they could envision a country with more farmers acting as true stewards of livestock and the land, farm families that can afford health insurance, orthodontia, and even that one Disney World excursion per the American Dream. Or maybe our cheeky name just made them snort. Either way, it was good.

A few weeks later, Jason found me tucked into our basement, writing my thesis. I could hear his obnoxious rock music ringtone accompanied by heavy steps. When I looked up, the cell phone was thrust in my face and he was pointing out the caller ID. It read *Mountain View, California*, the corporate home of Intuit. We giggled and put our heads close together so we could both hear the phone conversation.

Now it's one thing to say, "Imagine if our message of hens on pasture was on a national stage." It's quite another to be waiting, weak-kneed, in the wings when—holy *Gallus gallus domesticus*—LoLa made it to the Small Business Big Game Top Twenty.

I'd both craved and cursed the bright spotlight that was to be cast on Locally Laid. It was great for the business and even better for agriculture of the middle, but I was convinced that I'd say something stupid with my quirky sense of humor and offend swaths of folks. But fears or not, it was happening. I cranked out press releases before work and managed to land more television, this time coaxing a Twin Cities reporter 150 miles north to do a story. Our hometown newspaper gave it a column inch and a half, just below a story on the naming of the new monkeys at the zoo; the daily was likely sick of my fervent public relations campaigns.

In early November 2013, we got a call from Steve, an advance scout from Intuit. He was coming to find suitable locations for a professional photo shoot, which the company wanted of all the top twenty finalists. He'd need to see the farm and, of course, our office. He needed it the next day.

Hanging up, I looked from the couch, where worn throws were arranged like shootout victims, to Jason's desk encircled with a five-foot radius of paper splash. There was nothing I could do to make this look like the operational headquarters of a Super Bowl ad–worthy business. With the job, the thesis, the farm, the family, and now this contest, I'll admit we hadn't been winning any Good Housekeeping awards.

I knew what I needed to do: tumble-roll into Operation Create an Office, flipping our currently unrented basement apartment (walk-in junk drawer) into a contemporary workspace in under twenty-four hours. This included an emergency trip for throw pillows (in our company colors of orange and blue), new towels and cute soap for the adjacent bathroom, and pencil cups and all things office-y. Like a file cabinet.

There were some sizable bruises on my legs from hauling furniture around our house to fill the space. For our bare walls, our friends at our local UPS store bailed us out again. Beth and Jay took one of our farm photos and cranked out a gallery-wrapped canvas—on a Sunday.

Just as I finished hanging the huge chicken portrait, our UPS friend walked out the back door as Jason and the Intuit rep were walking in the front. I knew this because Milo was hollering, "Mother Bird has landed! I repeat, MOTHER BIRD has landed!"

Jason gawked around the new office like a tourist as I willed him to stop looking so bedazzled.

Steve, a young guy with a kind smile, gave the office a quick nod of approval and calmly explained what would happen next. Intuit representatives would be flying in the next day for the photo session, and these images would be later used for "marketing purposes." After a pleasant twenty minutes, he was gone.

Later, he texted us saying the Intuit folks wanted to go see the chickens before coming to the office.

We tossed in bed that night.

"Do you think it's like a job interview? Checking to see if we can handle making the next round?" Jason asked in the dark. Despite the voting, the software giant had the final say in which companies advanced to the final.

"That makes some sense. I'm sure they don't want to accidentally crush a small business with too much hype," I answered.

"We're pretty damn small," Jason said.

"Yeah, but we have office art."

The next day, we awoke to cold, intermittent fall showers— and spitting snow. I left Glensheen early to gather up the children from school to head out to Wrenshall, past the population sign that read *308*.

The pasture where our coops sit was heavy with clay, and when clay gets wet it turns into an especially hellish kind of sludge. It's dangerously slippery and really kicks up those natural farm smells. I hadn't worn my muck boots because I'd come directly from the mansion and it was just a few pictures, right? How long could this take? I quickly regretted it. Today's mud was the kind that would suck off a shoe and never return it.

Then we waited . . . and kept waiting. Minutes turned into hours and we jogged to keep warm. We pulled on every piece of castoff work clothing we kept in the farm shed. Still, the children were cold. I dug around until I found some cardboard to cover my vehicle's floorboards and entertained them in the car by taking silly selfies and singing to the radio, over the drone of the heater.

Finally, in my rearview mirror, I saw what looked like an entourage coming down the dirt road. I turned around to take in a big black Escalade, a Prius, a couple of sedans, and—a limo? Yes, Locally Laid was seeing its first limousine. Apparently, the crew had been staging at the Wrenshall convenience store. (I don't think they bought their cover story that they were filming a hunting video.)

When I say crew, I mean camera crew. Five or six men and women with hefty gear tumbled out, along with someone wielding an enormous boom mic. There were also really beautifully dressed Californians wearing brand-new rain boots. A few even wore bright red rain ponchos, still sporting the fold creases from the packaging. (The camera crew had no boots, and I'm pretty sure all of them had to throw out their sneakers after the event.)

And hey, isn't that the guy with the good teeth from TV? Yes, Bill Rancic, who won Donald Trump's first season of *The Apprentice*, was there. We hadn't really gotten into that show, but because of Milo's obsession with business, we'd watched that first season. A couple of years back Milo had dressed as Donald Trump for Halloween. He'd spent six months growing out his hair to make an enormous Trump-like comb-over. On the field that day, Milo tugged my sleeve, wanting me to ask Bill if he'd ever peed in Trump's gold toilet. I gave him a slight back-and-forth with my chin; I just couldn't ask. Honestly, I barely had the presence to form any words, and I was pretty sure *pee* and *gold toilet* weren't going to be among them.

Within seconds, we were ushered out to pasture, still wearing all our mix-matched clothes for warmth. Before we could grasp what was going on, a huge congratulatory banner was unfurled and every camera was trained on us for reaction. And we blinked, we stared, and then, one more blink.

Stunned and cold, we made terrible television. Even though they were telling us we were in the top four and I could actually read on the banner that we were in the top four, it just didn't make sense. I mean, we're us—a small farm in Wrenshall. We sold our first egg barely a year ago and were struggling to just make it work day to day. It was all I could do to not holler, "The office is fake! We didn't own a real file cabinet until yesterday! It's all a sham!"

Soon Bill was asking us questions, interviewer style, and the boom mic hovered over our heads as our reluctant lips remembered what to do. He asked about our farm-to-store model, our difficulties raising capital. I may have confessed to the whole world that I recently signed away my entire retirement fund, and that now every last dime we had was in the farm. I'm pretty sure

I talked about how if we won this commercial, during the most-watched televised event in the United States, we'd be taking Real Food out of some tucked-away section of your grocery and giving it the same stage as a Frito-Lay corn chip. Or a Dorito. Surely I'd offended a good number of people with that statement.

Then I saw Jason lead Bill and the camera crew over the solar fencing and inside a hoop coop. And while I was involved in polite conversation with some of our guests, I was distracted with what Jason was up to. He looked to be holding a chicken backward, football style, and from my peripheral vision I swear I could see him showing Bill, a beautiful and manicured man, how to assess a hen's fertility. I blinked, but the image of the reality TV star on our muddy, smelly pasture with his fingers on our bird's back end evaluating her pelvic bone remained. It was a Salvador Dali painting come to life.

I was nervous talking to the Intuit communication and marketing department members. They were holding hens with gusto and politely not mentioning the wafting odor: a combo featuring part bird, part manure over the sulfurous bloom of waterlogged clay soil. I turned just as one particularly well-dressed gentleman took a huge digger, like a cartoon; both feet went out from under him.

My mind was screaming, *Californian down!* but he piped up with an affable "I'm okay!"

Poor guy, I hoped he had a good dry cleaner for that gorgeous wool coat.

Oh, and as a side note: they never came to the office.

Chapter 19

Does waking at 2:45 a.m. qualify as getting up early or just sleeping poorly? I hadn't been setting an alarm during those blurry days of the commercial contest; rather, I'd been letting the tendrils of loose ends drag me out of my warm bed.

But what had popped open my eyes that particular early morning was an excitement hangover. My cell phone had started ringing early the day before at work, then nonstop with people clamoring to tell me that LoLa had a celebrity tweeting.

"Michael Pollan?" I'd said flatly into my cell from my workplace's standing desk, taking in Glensheen's view of snowy trails and bridges down to Lake Superior. "The one with the food book empire? The movie *Food, Inc.*? Who hangs with farm god *Joel Salatin*?" I noisily pelted down the stairs of the carriage house office to the communal printer, where my latest script for the mansion's Christmas tour waited for me. Pinching a vulnerable mint from a co-worker's candy dish, I went on, "You're telling me that Michael tweeted about our farm . . . in Wrenshall?

Can I call him Mike now? Does this mean we're friends? . . . No, I'm not mocking you, I'm mocking my own crazed existence."

My brain rejected this information, even when I went online and saw the less than 140 characters for myself. I mean, this is akin to having a big star, someone whose work you really admire—say, Jesus Christ—single you out of a baseball stadium full of fans and give you the nod, saying, "Hey, nice work."

To keep the excitement up, we planned a series of new videos. One of these clips had Jason speeding up the prairie on the four-wheeler, just as the gorgeous pink of predawn illuminated the chickens on pasture. Jason was honking the vehicle's tinny horn and then leapt out with a bullhorn shouting: "LoLa, big news! We're one step closer to the Super Bowl!" A couple dozen takes later, he got it.

Another featured a chicken "driving" our van, swerving down an empty country road, chased by police. She was pulled over for "tweeting and driving" about the Big Game. The Duluth Police Department was both generous and good sports, giving this silly clip all they had.

The new videos gave something fresh to be shared on social media and, I thought, would allow needed television B-roll, those images that flash onscreen while a reporter talks over it, for news reports. They could cover the story without having to trek out to the farm.

It was working.

I'm not sure if there's a technical definition of groundswell, but Duluth, during those three months, wowed us with its buildup. It was like watching a movie montage as the scrappy misfit underdogs gave it all they had against their well-favored rivals.

As Abbie and I drove through the commercial business area, we started to see the marquees. These signs, the ones with the changeable letters at theaters and car washes, even the big regional mall, now read *VoteLoLa.com*, listing the website Matt had cleverly set up for the online voting. A little bubble of something— Joy? Astonishment?—formed in my chest as we counted out over a dozen of these signs. Considering all those pitiful times on the prairie, this turnaround was stunning.

"Mom, you aren't crying, are you?" asked Abbie, a bit of tease in her voice.

I smiled through tears.

And these signs were just part of it. Thousands of small bookmark-sized notices about the contest were printed at cost, again by our UPS friends, then hot-glued onto takeout boxes from a half-dozen pizzerias. They were also placed in hotel rooms and given out with restaurant checks and grocery store receipts all over the city. Homemade signs popped up in storefronts, free ads blinked on websites, and lawn signs sprang up all over town. When we had a Get Out the Click rally, some twenty-five people stood with us on a bitterly cold morning to dance with signs for commuters. Curt, our business coach, even showed up in a chicken suit.

The VoteLoLa siege exploded online. Dozens of people changed their Facebook profile picture to our VoteLoLa.com graphic, and the social media giants at aimClear, when not busy with their own big-name clients, helped us gratis to scare up traffic online. A friend of our friend Linda donated a digital billboard—a freaking BILLBOARD. Our girl Whitney would gladly "Peep Your Ride," painting our logo and website on your rear windshield. Even the *Duluth News Tribune,* the paper that had once reported the competition in two sentences below the monkey-naming story, now thought it was front page news. Then they gave us a free print ad.

I set up a stint of television interviews for Jason in the Twin Cities, which he did while Brian farmed and I showed my face at work. Streaming some of the newscasts online, I watched Jason explain how by sourcing and selling locally we were strengthening rural economies. Then the reporter queried him about the ad agency that handled our creative. "Oh," Jason responded. "That's just a middle-aged woman in her jammies at five in the morning— my wife, Lucie!"

I shook my head, smiling and swearing simultaneously.

Later the marketing folks at Intuit would ask us what happened. Not about Jason's foot in mouth, but about our explosion of community. They hadn't seen any of the other regions get riled up quite the same way. It was hard to explain to people in prosperous and warm California, but I think Duluth saw in our little startup a mini version of themselves. The classic tortoise-not-favored-to-win story.

Our little port city is probably best known for three things: undisputed beauty, being slapped across the weather map with inclement conditions, and its postindustrial economic apocalypse. In the 1980s, it once sported the billboard, "Will the last one leaving Duluth please turn out the light?"

I can only guess that this community saw something to rally around. And it was a chicken.

While our region clicked their love, the daily vote I was most touched by was the one jointly cast in Maine. My mother's dial-up Internet couldn't handle the graphics-heavy Intuit site, so my octogenarian father drove her the mile to the Winslow Public Library—every single day. There she fulfilled her daily e-ballot on the public terminals. My father didn't vote himself, but this didn't bother me. He's not a computer guy. But his ensuring that my mother got there was a much-appreciated act of support.

And we needed that support.

For me, it wasn't so much about winning the commercial. Don't get me wrong, that's big, huge. But I was far more interested in the media around the winning of the TV spot, the dozens of national television, radio, and print outlets that would interview us the morning after it aired. The idea of sitting on the *Today* show couch in front of busy Americans wearing hot rollers or running on basement treadmills, asking them to question, for just one moment, where their food comes from, well, it nearly stopped my heart. Even if some minuscule percentage of mainstream America, for the first time, asked their grocery managers to source local products, then everything we had gone through and indeed were going through had reason.

I understood that lofty goals required sacrifice. Earlier in the week, I'd walked into the house with take-out food and saw cereal and snack food boxes littering our countertops along with an assortment of dirty Fiestaware. All the pressure and clutter worked on me, and I know because I was losing track of important items. My natural reaction to stress is to emulate a shaken soda bottle, bursting forth all objects of value across my scattered world—my wallet with credit cards and driver's license, my favorite bra, and all matching socks. Even Jason, for the first time in his life, mislaid a set of car keys. That he was out of town, requiring a tow truck, only added to the excitement.

Fortunately, we had guidance. Once a week before work, a small group of community leaders and businesspeople gathered around a conference table to talk contest strategy. These busy folks freely gave their time and talent and abused all their personal connections on Locally Laid's behalf. Truly, it pained me.

I carried an oversized feeling of indebtedness and, like a kid with too much candy, it felt like an overindulgence of too much nice. I had a bellyache from all that sweet. After our last meeting, as everyone scurried off late because of us, it hit me exactly how much people were doing. All that generosity and faith spiraled in my head and chest and I realized that, more than anything, I owed them victory. Not just for them but for this city, this region, the whole dang agricultural movement.

Walking out to the parking ramp, I inhaled sharply, bent at the waist, and vomited across the concrete floor. Now, at the age of forty-three, trying to wipe gag splash off my pant legs with a glove-compartment napkin, I truly understood the adage: "It is better to give than receive."

I had no idea how I could ever repay these people.

Other than by winning.

We did not win.

We learned this via a thirty-second phone call from the contest's marketing department, a few weeks before we were to show up in New York City for a huge football-themed celebration. And, to make matters worse, we were sworn to secrecy with a binding nondisclosure form.

Jason was stunned, in the literal open-mouth, cannot-speak sense of the word.

After sprinting through those crazed months, it was like we'd hit a wall and Jason, dazed from the impact, couldn't quite fathom that we'd lost. Then again, he is simply a more competitive human than I am. In school sports, I remember passing off field hockey balls with the clear thought that this little rubber orb was going to make my opponent so much happier than it ever would me. Take it, enjoy.

This is not how Jason is built.

It was more than his sheer want for victory. In this contest, Jason and many others felt that the winning business, a well-funded toy company, had used questionable tactics. Some said illegal ones. They'd created a high-end promotional video (with a budget more than our annual revenue) and dubbed it with well-known copyrighted music. And when the band asked them to take it down, the toy company preemptively sued them.

The misstep, or stunt, depending on your perspective, earned them several days' worth of national news cycles and lit up social media sites like Upworthy and BuzzFeed. It was the subject of talk shows; it engaged people. And like the saying goes, all press is good press.

From that standpoint, I admired their ingenuity. I had to. I asked Jason, given that this prize was of astronomical value, whether he really thought eggs could compete against the allure of plastic playthings?

I resigned myself to second place, which offered a great prize package of its own—including a Hollywood-made TV spot that aired on cable for six weeks. But Jason was not done. He called right back, asking for a recount, reconsideration, restitution. Intuit was gracious, even assigning us a big-hearted staff attorney, Lisa, who served as part legal interpreter, part therapist, and followed up on each of his concerns. And though we were given no indication that the outcome would change, Jason held on to a thin hope that somehow Locally Laid would pull it off in the end.

Given all that our community had done, it had to change, right? If this was the movie it felt like, then in this crazed drama the scrappy little egg company and its oft chicken-shit-covered owners trying to save the world from industrialized foods would, against all odds, enjoy an upset victory against the bigger corporation selling toys made in China. At least, this was what Jason thought.

I tried to press reality into his head, but he continued to believe. Jason believed all the way to the announcement in New York City, where he arrived late—trapped at the annual International Poultry Conference in Atlanta, a city completely flummoxed by snow. He was too late to join me on the planned media junket.

Truly, it was a bizarre day in New York City. It started with an interview at Fox News Business—the national one. And I had to do it without my favorite farmer. Arriving at the forty-five-story skyscraper in Manhattan, I emerged from our small fleet of black Escalades (itself a bit unnerving). That was when I started second-guessing my outfit. I'd ironed my now broken-in Dickies jacket for the TV spot and was pretty convinced I was wearing the only nonblack garment in a six-block radius of Times Square.

I'm also sure I came off too bubbly. Answering the questions of the tiny newswoman (she was likely a dress size zero, a garment dimension rarely seen in our corn-fed corner of the Midwest), I fell into my overcaffeinated, default TV persona that conveys how dang HAPPY I am to talk about chickens.

You see the birds exercise! And eat grass! And lay eggs with less fat! Praise be and hallelujah!

It was the opposite of deer in the headlights; it was deer charging the glowing orbs and leaping over the hood to clear the roof of that oncoming sedan. (I suppose there are worse things, but really, would a little poise kill me?)

Very quickly it was all over, and my oversized public persona was stuffed back into the fleet car. It was nearly time for the big public announcement that would roll out on the Associated Press newswire. But first, Matt, the handsome man who'd fallen in the farm muck back in Wrenshall, bought me a whiskey in the hotel bar and toasted Locally Laid. By the way, the wool coat he'd soiled at the farm was forever ruined, though he'd kept it as a memento of a great day.

I kicked back my drink, coughed, and excused myself to make a call.

"The announcement is going to roll out any moment," I said to Jason. There would be no last-minute reprieve. No big chicken in the sky to save the day. I could hear the bustle of the overrun airport through the phone's receiver, along with his monstrous sigh. Jason had hoped for a last-moment turn of fate. It was not coming.

"Yeah, I figured." There was a long pause, the gathering of momentum. "It just could have been so cool, Lu. I mean, what this TV spot could have done for something important, like really important." His voice trailed off.

"Honey, we've done so well. Like statistically, improbably well."

He was quiet.

I wanted to pull him through the phone line, envelop his big body with my smaller self, and comfort this all away. Just as I yearned to shake him by the shoulders and awaken in him all we'd had been given, by the community, by Intuit, by *life*.

There was a lot to be grateful for. After all our megaphone shouting about the existence of middle agriculture and the glory of chickens on pasture, a consortium of lenders had recently rubber-stamped our loan application. Certainly, all the publicity and the increased sales it generated helped. We were finally going to buy a farm of our own.

Owners Dean and Sandy even allowed the birds to be moved

in before the official close. On the night before Thanksgiving, Brian, Jason, and our architect friend John slipped headlamps over their wool caps. Transporting the chickens at night was the least stressful on the birds, though with the pelting snow, seen in the illuminated circles of the men's lights, it wasn't the best for the humans involved. They wrangled each sleeping hen off her prairie hoop coop roost, crated her, and drove her to the new barn. There were many trips in the snowstorm that dropped some five inches that night.

And while the new barn was imperfect (later we would learn just how imperfect, when we saw our winter heating bill), it gave our girls a warm shelter. There would be no more frozen waterers or frozen eggs or frozen workers this winter. Not long after, the temperature sank into a record streak with several days below minus twenty. It's hard to imagine the flock could have survived that abnormal polar vortex freeze under tarps.

As we walked the barn later, the chickens looked healthy, lively even, and the space smelled surprisingly sweet with fresh straw on the floor. Of course, even this much-improved setup would have its problems and we'd eventually go back to ladies in hoop coops for the summer to achieve better pasture rotation. But at that moment, we were ecstatic that these chickens were out of the elements.

The new location also had a space we could build out into a new egg-washing room. We didn't know it then, but the startup grocery store we'd been renting space from would be out of business within a year. Having our own land that we could build on was more important than ever.

But I didn't need to think about washing eggs right then. Up in my posh Gramercy Plaza hotel suite, a place so dark and plush one risked being swallowed whole by its upholstery, I waited for the contest announcement to scroll across the newswire. I repeatedly tapped refresh on my old Mac, like an old-fashioned drinking bird toy, because I needed to be the first to congratulate our competition. It was important that I set a positive tone on the social media sphere. Knowing our supporters would be wholly disappointed, I wanted to tamp down any possibility of virtual rock throwing at the winning company and certainly not at the contest sponsor Intuit.

Also, I'd been penning a blog most of the afternoon, a love letter to our volunteers, and wanted to get that online right away. In it, I stressed that because of their enthusiasm, a sustainable startup got serious national attention. Media impressions for Locally Laid were cresting into the billions. A whole lot of folks who had never heard the term *pasture-raised* now had an inkling that exercising chickens on grass was a good thing.

Before leaving for New York, I'd talked to my favorite editor and friend, Eric, about losing and how crestfallen LoLa's ardent peeps would be.

What he said next, in his rich radio voice, was a lesson culled directly from *Charlotte's Web.*

Now it may have been a while since you've read the E. B. White classic, but you'll likely recall that Charlotte, a clever spider, spun words into her web in hopes of gaining attention for Wilbur, her pig friend slated for bacon. This web writing causes quite the hullabaloo.

But—there's a bigger pig named Uncle. As the blue ribbon prize is based on weight, Uncle wins the blue ribbon. And here's the important part: Uncle deserves the blue ribbon. Just as the fair-going crowd is poised to disperse, the announcer calls another award, an unprecedented award—one for a special pig. Though Wilbur was not the biggest, he touched hearts and inspired wonder.

Many good things happened out of the contest: a community came together, farmers interested in producing eggs the way we do contacted us, and though ambitious Jason still smarted at our loss, I think we ended up right where we should be.

Still the underchicken.

The tiny plane home to Duluth was delayed because of poor weather, exceptionally poor weather. It was sleeting and they elected to de-ice the plane—twice.

"They should cancel this flight," I said, forehead against the ice-streaked window overlooking the tarmac.

I used to love to fly, but the older (and more mortal) I've become, the less exhilarating and the more terrifying it feels. Moments after the plane took off, there was a bing from the sound system and the captain announced that we were in for a bumpy ride.

He did not lie.

Our aircraft dipped and tossed, causing a group of women in front of us to occasionally let slip a little scream as I squeezed Jason's hand. Then I put my head to my knees and started to cry.

Weeping grew to sobbing and, while I was frightened as our small craft pitched in the gale, I felt a release of tension—over the contest, emotionally tending Jason, buying the farm, finishing my thesis, even the new Christmas tour at Glensheen. I just cried it all out—in a loud and unladylike fashion.

By the time we deplaned, I was composed and took the stares from other passengers in stride. I was officially the crazy crying woman on the flight. I did not care; it was the most relaxed I'd felt in months.

As we rounded the corner out of security, there was a wall of people. Despite the late hour and below-zero temps, a gathering of a dozen or so of our closest friends were there to welcome us. I gasped in surprise, and it was Jason's turn to get teary.

We were home.

ACT 5

Fledged

VERB: to have acquired wing feathers
large enough for flight

Chapter 20

When I was a freelance writer in Minneapolis, there were editors I'd worked with for years, with offices just across town, whom I'd never met. I'm not agoraphobic, and they were not unfriendly; it simply wasn't necessary. While I could easily recognize their electronic signature in my mailbox, I couldn't have picked them out of a lineup. This was perfectly acceptable and undeniably efficient, if not particularly cordial.

But it's not the way the Amish work. At least, not the way of the particular community we were driving to see that early spring day in 2014. These Amish folks are more on the observant end of the religious spectrum. Having no electronics in their home, they'll send us postcards to set a plan to call us from their community phone.

So, everything takes more time and patience, especially the building of relationships. It requires meeting you, shaking hands, and sharing a meal. It would not be confused with speed dating. Nor should it be viewed as optional.

As with all religions, there are those who are to the letter and others more, say, loosey-goosey. Take Lamar, our Amish partner in Indiana, just outside Chicago. He has a smartphone, texts, and is in on the Locally Laid joke. He even has a LinkedIn account (no picture, only the default silhouette photo place-holder, but what I'd give to see that outline sporting the wide brim of an Amish hat). He is clearly as progressive as an Amish guy gets, sitting on the other end of the spectrum from the peo-ple we'd be visiting today.

And this had me a tinge sweaty, for a couple of reasons.

For one, until recently, everything I'd known about the Amish culture was based on the 1985 movie *Witness*. My biggest takeaway: they would be immune to my self-effacing, pop cul-ture jokes.

And while these Amish folks were eager to work with us, they'd worked with smaller numbers of layers in the past; egg production at this commercial scale would be new to them.

The other two farmers we'd been partnering with were long-time chicken keepers who only needed to shift their operations to birds seasonally out of doors, be educated on pasture manage-ment, and maybe add some poultry-life enhancements to their barns like dust baths, roosts, and winter hay. But for this commu-nity, there would be lots to learn.

Of course, they'd be starting with more practical knowledge than we ever had (or will), and they're close enough for Jason to check in often. But what they had going for them was location. Their farm is situated nearly equidistant between two sizable egg-buying markets, Duluth and the Twin Cities area.

Even so, I hadn't been entirely convinced this was a venture we should take on. We had our own hens, a partner in Iowa, and one in Indiana. I felt stretched.

Then Jason, likely in an attempt to get me on board, showed me the area's demographics. The 2000 census reported a per capita income of just over thirteen thousand dollars, and the 2010 data showed a population loss of over 29 percent. Clearly, this was a rural location that could use the income.

Looking out at the flat, snow-splotched landscape from the van's passenger seat, I said, "I hope we have stuff to talk about."

"Well, they're making us lunch, so you can talk about that."

"What? We didn't bring anything! We cannot show up empty-handed for a meal."

Jason shrugged. Clearly this fell into his large mental file labeled *Not gonna worry about it.*

So I worried for the both of us.

Before setting off, I'd done a little homework, dispelling a few myths. Like, I used to think that Mennonites were break-away Amish—folks who embraced the orthodox traditions but were more willing to interact with the modern world. Wrong. The tome of a book *The Amish* explains that both sects come from the Anabaptist tradition of sixteenth-century Europe. During the tumultuous time when reformers such as Luther and Calvin were stirring up radical religious change, some felt their stands were not going far enough. Enter stage left a faction known as the Mennonites, so named for the inspirational writings of former Catholic priest Menno Simmons.

But in 1680, a charismatic tailor and convert named Jacob Ammann came into the picture saying that Mennonites, while on the right track, simply weren't radical enough. He proposed many reforms, including all-but-complete exile from the outside world. This ultimately led to a schism in 1694 between those who followed the teachings of Menno Simmons, the Mennonites, and those who adhered to the more orthodox words of Jacob Ammann, known as the Amish.

You might think this austere lifestyle, with restrictions on nearly every aspect of one's personal freedoms, would be a dying one. And you'd be as mistaken as I was. According to the Rural Sociological Society, the Amish religion is one of the fastest growing denominations on our continent, expected to break a million folks sometime midcentury. With growth like that, a community outgrows itself every twenty-five days, sending forth representatives to buy more land and start anew.

Usually, after I study something, I am more at ease, more confident in the topic. But turning off the highway and seeing buggy tracks on the waning snow got my anxiety twirling. And although I dressed conservatively, what I did not think through was my hair.

The Amish, who also call themselves the Plain People, eschew vanity and adhere to a literal translation of the Corinthians verses that straight-up tell women to cover their heads. But I'm embarrassed to say that when I first met our hosts my first thought was how unattractive a big-nosed gal like me would look in a hair-hiding, starched bonnet. Now let me first say that Ruth, the woman of the house, has fine features and sports the cap quite attractively. I'm pretty sure I'd be mistaken for a male wreathed hornbill or perhaps a toucan.

I nose-compensate with big hair. It's not like I fuss with it, just wash and rumple it in my palms with whatever hair product is on sale. But given the recent snowmelt, the air was wet and tress-inflative. Like country-singer large or Sarah Jessica Parker full. Walking in the door of their comfortably modest home, I was immediately self-conscious. As we spoke our awkward hellos, I found myself attempting to iron my hair to my head using a flat palm. It sprang back up.

Ruth, perhaps in her early fifties, wearing a cotton apron over her dark homemade dress, didn't seem to notice. Truthfully, she seemed more at ease than I was. She introduced me to some of her older children; the younger ones were at community school, from which they will graduate at the end of eighth grade, and a beautiful baby granddaughter, some six months old. In her simple dress and little bonnet, she looked like a doll, perfect with fat baby wrists.

Sitting in their living room, we four adults worked through the details of an egg production contract and, specifically, the regulatory aspects of egg processing. We, more than anyone, knew that getting these washing facilities right is important. There are always hurdles, but they're additionally higher when working out ways to heat to exact temperatures and keeping clean eggs at a precise forty-five degrees without electricity. It's doable, just not easy. And our conversation circled around gas generators, woodstoves, and battery-operated thermometers.

There were times when Ruth's husband, Marlin, dug in on

points, flatly stating they did not require electricity for some small task, like candling. It seemed like a tiny issue, but I realized that this entire culture plods forward in our modern world only by stringently holding the outside world at bay. With both arms locked.

At one point during our conversation, Marlin, a serious-looking man wearing suspenders pinned to wool pants and a vigorous salt-and-pepper beard, turned his attention from Jason to me and asked, pointedly, "How old are you?"

I was wobbled by the question but also attracted to its directness. He asked in the subtle accent that flavors Marlin's English. To my ear, his *you* registers as *yew*, in deference to the Low German he also speaks.

I was silent for a beat.

"Did I offend you? I don't want to offend you." I could see he sincerely meant it.

"No, no, it's fine," I demurred, and answered that I was forty-three (though sometimes in poor lighting I can be mistaken for younger). He eyed me some more and nodded. I could only guess he was curious about us, too.

The men continued to talk numbers as I leaned into their circle of conversation. Jason, figuring a percentage in his head, was quickly corrected by Marlin. He may not have gone to high school, but Marlin was fiercely bright and used modern business terms like *distributors* and *margins* and *capital costs*.

"Could you bring in your computer, Jason?" he asked.

We'd left our Macs in the car, not wanting to disrespect their home with technology, but this is a good example of what we've found working with Amish families. They're happy to benefit from technology, just not to wield it themselves.

This can be surprising, and I defend the Amish from those who mock their acceptance of car rides or the use of generators or the computer chip accuracy of our calculators. Aren't we all just a big swirly of compromises? I think this as I write a check for produce from my local CSA and then run my debit card for toilet paper (and other items of excretion management like tissues and feminine hygiene products) at the megawarehouse store.

Jason and I hoped to hammer out a deal today and sidestep a few issues we've had in the past. We were (and are) learning when it comes to these partner contracts. There's little out there to model on, especially as we try to respect the autonomy of the producers. While we require our partners to raise eggs like Locally Laid does, we get that farming is one of the last domains of the cowboy. Farmers are disinclined to micromanagement. We build in as much autonomy as possible. As in, here are the standards in regards to treatment of chickens, the rules set by regulators about handling and temperatures, and the ingredients for feed to be procured as locally as possible—find your way to do it.

But there can be no Wild West in egg washing. One of our partner farmers had some initial trouble, skewing more earthy when it came to shell cleanliness. And now that we work with chickens every day, I get it. The eggs I bring home to my own family often have mystery flecks one can scrape off with a fingernail or maybe a feather in the carton. That's okay for us but completely unacceptable under the all-exposing fluorescence of supermarket lighting.

Commercial eggs must be pristine, sprinkled only by holy water, washed by the hair of organically certified virgins, with an ethereal chorus of angels singing as the carton is cracked open. This has led to awkward conversations with Amish producers and we try hard to not make it about them. Instead we draw the onus back to the consumer: typically a higher-income, educated suburban gal. And while she may appreciate the farm and theoretically loves chickens, she doesn't need or want to experience the barn's coarse reality.

It is hard for me to emphasize enough to our Amish partners the impact of a dirty egg out in the world—how an irate consumer may take a photo, put it out in social media, and it's there forever.

"And this, this would not be good for Locally Laid and all of us who produce for the brand," I say.

My speech is often met with silence.

It's nearly impossible to explain the power of Facebook or Instagram and their flow of personal anecdotes and goofy snapshots to people who have never themselves been photographed. It's like Jason once joked about the Amish: great at farming, bad at Twitter.

Ruth and her adult daughter, standing shyly behind her mother, called us to lunch. It was a meal served next to the wood-fire cooking stove, not unlike the historic one sitting in Glensheen, all scrolled iron legs and heft. The difference—this one was blazing hot. Ruth served a casserole meal of beef in a white sauce, green beans canned last season, bread, and a fruit crumble dessert. It was good and appropriately heavy for people who work out of doors. As I ate, I saw that nearly everything on this long wooden table was produced right there.

I sat near Ruth, who affably inquired about my projects, things I might be sewing, jams I'd put up last summer, and breads I bake. I had nothing to offer, to the point that as the conversation plodded on, I began to wonder, *What* did *I do all day?* Watching her desperately trying to create polite conversation, earnestly attempting to find something, *anything* I can do that she'd be able to remark upon was charmingly awkward. I didn't know how to tell a woman who'd recently helped a neighbor birth a baby that I nearly failed junior high home economics. That indeed my mother had to completely remake my gym duffel bag project, a sewing task that involved three seams and a zipper, and that I was banned from handling the school kitchen's plastic wrap given my ruinous attempts to start the roll.

What we had in common is motherhood. I complimented her children, who joked quietly with each other at the other end of the table and, once the dishes were done, alternately entertained the baby and read books. I noted that one of the teen boys studied a music book that looked like hymns, but I couldn't be sure given they weren't in English. I was taken that a boy his age would be interested in music like this.

I later read that singing is the courtship ritual of the Amish young. While it varies from community to community, it's one of the culture's oldest activities, going back to the 1800s. It's basically a young folks' choir, but seemingly a little romance happens with the chatting and laughter in between songs and, of course, the buggy rides back home.

Learning this made me smile. No matter how different we were, there are some basic universal tendencies. Boys will do what it takes to meet girls.

Months later, on a different visit, an evening one, we sat around that table again with the woodstove blazing to boil water for tea. The younger daughters were there this time, sitting beyond the table, bonneted heads together in the ethereal light cast by the lantern. They whispered, giggled, and sneaked glances at me. I thought they looked remarkably like Abbie. I cannot imagine what I looked like to them.

This encounter felt more relaxed, or maybe I was just more relaxed. There was laughter as Marlin recounted one of the big distributor semis coming to pick up eggs, overshooting the driveway, and getting terribly stuck.

"Well, we hooked up our team and just straightened them on out," Marlin said. He was smiling. I can't imagine how fun it is for the Amish to get one over on the "English" (as they call us) and our technology. Even I felt a flush of pride on their behalf. I suspect the truck driver made no mention of the incident back at the home office.

As we discussed some accounting issues and ideas to make the payment system better for the producers, I noted that Ruth was not sitting at the table, only the men—her older boys and me. Nibbling cookies, I felt that I'd screwed up.

Getting up to make our good-byes, I gathered up mugs and brought them back to the kitchen where Ruth stood near the sink. I wanted to apologize for not sitting with her.

But she jumped in first. "You shouldn't have taken your shoes off, it's too cold in here for that." She looked down at my stocking feet, filled with chilly toes. "We're insulating this whole part of the house soon. It'll be so much more comfortable then."

She said this with a tinge of pride. I smiled, thinking this home improvement was likely linked to their new egg revenue. Though communication would never be easy with the Amish, I decided then not to worry about where I sat. No matter which chair I took, I could see Middle Ag at work.

Chapter 21

"Can't you let me be happy for more than seven goddamn minutes?" This came out louder than I intended. I rubbed my hands, then my temples.

In the summer of 2014, Jason and I were making the drive down to Minneapolis for a speech practice session. We'd won a spot in the finals of the Minnesota Cup, a business competition that ferrets out innovative ideas with a thirty-thousand-dollar cash prize. As I was going to be the one presenting, I was already touchy. There was some pressure. Although the farm was technically making money and most everyone in the region knew our company name, nearly all profit was funneled back into the business for improvements, debt service, and to pay our small part-time staff.

That was when Jason said, "There's really no good time to tell you this." I snapped my head up from my notes. Dean and Sandy, he said, the former owners of the farm, who had stayed on as great tenants, had given notice.

My mouth fell open and I gasped, reacting like a bucket of water had been upended over me.

"They've inherited a house on a lake." Jason spoke slowly and gently, like he was trying hard not to spook me. I recognized it as the tone he uses when prattling to a hen he's about to snatch up in the field.

"They're looking to make the leap in the next few months, so it would be a good time for us to look at moving out there."

I don't think my allusion to seven minutes was an accident.

A psychologist I know claims it takes seven years for a place to feel like home. And that feels about right. In that spring of 2014, we'd nearly summited that number and Duluth felt like a place we were from. Doing errands now meant seeing people we knew, we got the inside jokes and the political asides, and even the gray skies now felt somewhat sheltering rather than cause for wide-arcing mood swings. Or biting.

Plus, the past few weeks had been happy for me, ecstatic even. I'd given notice at Glensheen, leaving to both work for Locally Laid and finish my book, which had been recently purchased. I was thrilled on a number of levels, but especially that I'd be more available to the kids. Be on top of their everyday school needs. Make eye contact. Increase the odds they might pick me out of a lineup of other middle-aged women purporting to be their mother.

Also, I was looking forward to getting back to some of the other things I'd missed, like sleeping past four a.m.

So, of course, Jason wanted to move.

"Don't Dean and Sandy's kids go to school in Wrenshall?

I . . . I thought they were going to stay until they graduated." As I stammered through the last part, I remember hoping that was the case. It was unlikely that anyone had ever said that.

"I don't know," said Jason. "I just know they've given notice."

Certainly, the farm is pretty, in a country-living sort of way. And while I grew up with plenty of land around my parents' home and even some round hay bales on its few front acres, Maine rural feels different than the Midwestern version. Land in topmost New England is hemmed in safe by far-off hills and forests of thick pine and birch, some 17.7 million acres of woods.

In Minnesota, rural means open, sweeping—exposed. And to me, that's unnerving.

Back when I first drove to the middle of the continent in 1995, I remember listening with increasing intensity to a book on tape, avoiding eye contact with the ever-widening sky outside my windshield. When I looked too long over my steering wheel, the broad vista made my heart rate quicken and my breathing labored as I envisioned scenarios of breaking down, fully vulnerable to the elements.

That was when I realized how much I like trees. While I'm no Maine Guide, the woods have always felt like refuge—a place one could build a shelter, find shade. Where dew would reliably gather in the predawn morning. But the sheer expansiveness of the midcontinent sky terrorized me. Though I wouldn't have put it this way back then, it created the feeling of a chicken without cover, waiting to be picked off by a hawk.

I call this IFP: Irrational Fear of Prairie.

"But the children are in school in Duluth," I protested. Abbie would be entering high school next year, and though I'd desperately wanted to bubble-wrap Milo for the start of middle school, it'd actually been a fine social experience for him.

"They wouldn't necessarily need to transfer; you could drive them," he said.

"Drive them? The Duluth schools are a good forty minutes one way. And then back home again? And what about swim practice? And Milo's Dungeons and Dragons night? He wants to be a ski cadet this year at Chester, you know."

"You could go the back way, through the state park. It's a gorgeous drive," he said.

I stared daggers at him.

He went on. "Lu, you're not seeing the positives of the children growing up at the farm. They can learn real skills."

I could almost see it playing out in his mind, visions of our kids helping him rip apart tractors and gathering eggs. In my own head, I did some quick math. I could easily be driving some four hundred miles a week, leaving the house at six thirty in the morning and not getting back till after eight at night. I wondered how pretty the drive was in the dark.

"But they won't be at the farm, and neither would I. We'd be in the car."

"Lu, it'll work out, it always does," Jason said.

It always does, I thought, *because I jump on the parenting grenade, containing the shrapnel of chaos.*

"Jason, I just got my degree and can start applying for teaching jobs now. Two colleges are walking distance from our house," I said, clumsily gesturing toward Duluth behind us. "And another two are only a fifteen-minute drive—and you want to move now?"

He shrugged.

After so many years together, he knew I needed to mull. He, on the other hand, was mentally moving walls in the new place. I recognize that change energizes him, gives him a bounce, whereas the prospect leaves me pining to see through the life I'm presently living.

I spent the next few days in all shades of unnerve and anger.

The next week, we convened a family meeting, making the traditional pro and con columns. Milo, now eleven, was wide-eyed and sat stiffly at our dining room bench, having caught my aversion to all things new. Abbie, at thirteen, seemed pliable, curious about the possibilities.

After nearly an hour of discussion, the list tallies were nearly equal, even after heavily weighting the category of "time spent in car" with its due oppression. And as far as gut reactions go, we four were evenly divided—Jason and Abbie were keen for adventure, as Milo and I clung to the status quo as to masts on the proverbial sinking ship.

Trying to be all adulty on this issue, I agreed to walk through the farmhouse. Look at it with new eyes. That Saturday morning was warm, and the sounds of our chickens' gentle clucking percolated over the landscape. It was clear. It was verdant. It was a setup.

It was over a year back that I'd done a quick winter walk-through of the home when we were purchasing the farm and, while I'd seen it from the outside plenty of times, my recollection of its details were foggy ones. For one, I hadn't remembered it was built in 1998, a virile teen compared to our current geriatric Duluth house, leaning on its walker as it turned 100 years old. This gave the farmhouse some pluses, like real insulation and decent windows. And as we walked through the house, I couldn't help but see us enjoying the bigger living space.

Jason came up behind me and pointed to a corner. "This is where I think we should have a woodstove."

I sighed deeply. Growing up with a woodstove in Maine, I can tell you there's no heat like it. As ludicrous as it sounds, it's just warmer. I've pined for that through-the-bone warmth all of my adult life. When that cast-iron door is opened and one can see the fire behind the metal mesh, there's nothing like the mind-emptying gambol of flames. It's humankind's original hypnotic screen.

Our friend Mike, a gifted carpenter, had joined us and was making suggestions on opening up the space, filling it with light. Mimi was along, too, dizzy at the prospect of getting us out of our tiny home into something bigger, better.

There were advantages. The kitchen had a much better lay-out than our current galley one and the basement was largely finished, complete with a sauna, and could easily be a large family entertainment room or teen hangout.

Then we came to one of the bathrooms. Inside was a soaker tub. With jets.

"Oh God," I said. The house was nicer than I remembered, than I hoped, truthfully.

"Moving here just seems the natural thing to do," Jason said.

He was right. Residing in the city of Duluth while being farmers was like a story that didn't quite fit together. When people learned that we didn't live with our birds, you could almost see the cognitive dissonance pinging in their heads. Like we were cheating.

I juggled some ideas. One was to rent out the upstairs of our Duluth house but leave the teeny basement apartment for the kids and me to live in during the school week. It's a one-bedroom with a second bed in the combo living room/kitchen. Very small, but doable—maybe. This way, the children could stay in their sports and clubs, while I taught.

Fortunately, it was taking the tenants longer to ready their new home, so I was granted more time to make friends with the idea.

And I had to go visit my parents in Maine.

I drove my rental car through rural Maine towns best known for their proximity to somewhere else. I was spending part of my trip home in farm country because of a book, *Food and the Mid-Level Farm*. It's a collection of journal articles on the struggles of farms like ours, and I wanted to talk to the project coordinators. Going down the line of them, I learned that one had died, one was too ill for visitors, and another didn't return my calls. Toward the end of the list was a Dr. Stewart Smith. He was from my alma mater, the University of Maine, and happened to farm just a couple of hours from my parents' home. Best of all, his wife said I could come.

After stopping twice for directions to fields, I was pointed to a set of dirt tracks that led into a field off the aptly named Old Stage Coach Road. My teeny energy-efficient rental car, which seemed utterly adorable at the Portland airport, was now absurd. Its miniature tires heaved along this washboard path, long grasses tickling its low-riding frame. I gunned the four cylinders through the wet, low spots, counting on the forgiving attribute of momentum, and arrived at a cultivated field, a hundred or so acres, tightly hemmed in by scrubby trees and thin pines. Farmers never forget that Maine is 90 percent forested.

Pulling in past a running refrigeration truck, which I'd later learn costs a hundred bucks a day to cool freshly picked veggies, I saw people. I could see from the gaggle of ten or so workers, bent at their waists over rows of squash and cabbage, that they did a lot of work by hand.

Soon a rugged, compact man approached wearing a Red Sox cap and striding with the confidence of a crew leader. Adam informed me that Stew wasn't there but would probably be out in about an hour. I was invited to wait. Just as we were finishing our conversation, a pickup truck pulled up with a couple of workers. The driver, a big guy, engaged us in a friendly conversation about fishing in which Adam let slip he'd recently caught a bass, "a three-poundah."

I smiled.

Whenever I return home, the New England accent jumps out at me now as surely as my big roundy *O*'s (as in Minn-Oh-Sota) cause eyebrows to arch, often inviting the less-than-polite query, "You from away?"

After assigning picking tasks to the workers, Adam walked the rows with me looking at the oversized zucchini plants. "We farm conventional in this field, growing a variety of vegetables," he said, adjusting his hat on his short brown hair. The website lists some twenty-five types of produce, some organic certified, some not—including the cucumbers and summer squash the crew was working on today under the hot sun. The model for this operation is to grow a wide variety, so they can bring lots of different produce to fewer stores, handling the distribution themselves. This differs from the commodity model, which specializes in what a farm's really good at, like carrots, forcing the farm to work with a number of distributors to get the produce out farther. One store can take just so many carrots.

It wasn't long before a pickup pulled up. A trim man with white hair and dressed in neat cotton work apparel got out and shook my hand. Stew had grown up in central Maine back when there were some twenty-four farms in the region, including his family's place. By 1977, there were six. And these were no longer diversified farms with a variety of harvest and livestock, but rather simplified, mono-crop operations, growing potatoes for chips, on an industrial scale.

Stewart went to Yale and later became Dr. Smith at the University of Connecticut, earning a PhD in economics that started a twenty-year career as a professor of sustainable agriculture policy at UMaine.

I didn't know it then, but he'd also served in a bevy of other positions, including agricultural expert to the state's governor and even President Jimmy Carter. Stewart was involved in a list of institutions the length of my arm. Being underinformed of these accomplishments was a blessing, as it likely would've triggered more nervous prattle. I was already sheepish about bothering any farmer, much less one who moonlights as a presidential advisor.

He invited me to sit in the truck, out of the hot sun. The cab reminded me of my father's work pickup back when he was a home builder, a big American model covered with a layer of dust thick enough to write one's name in. When we'd settled a bit, Stew looked at me and smiled with an expression that might be interpreted as the polite inquiry, "So, why are we here?"

I rambled a bit about my admiration for his book and the threads of similarities between our operations and then finally blurted out, "I'm just hoping you can help me understand why agriculture is so . . . *broken?*"

He exhaled. "Well, that's because agriculture is regulated by neoclassical theories of economics and it adheres to the first law of thermodynamics and not the second law of thermodynamics."

At that moment, Stew might as well have been talking to a house cat.

For the next hour, the doctor talked while my hand careened across my notebook scrawling a collection of scribbles that would puzzle me for months to come. I knew I'd stumbled upon something important to my understanding of agribusiness—a pitchfork-shaped Holy Grail—but could I wield it without poking out an eye?

So began the second phase of my education. This conversation led to more academic journals, and even an economic theory book appeared on my nightstand. As a woman of average intelligence, I'm not going to pretend any of this came easily to me. It pushed my limits, even prompting a piteous decree from my side of the bed, book light clipped to a journal, highlighter in hand: "Jason, if I don't read a vampire sex novel soon, I might not make it."

So, to save you such desolation, I'll just give you the abridged version.

There are two camps of economic theory. And let's be clear. They are each stalwart in their thinking and not prone to mixers.

First, there's the First Law of Thermodynamics tent, a crowded and popular place given that its tenants are the underpinning of modern neoclassical theories of economics. These are the cool kids. Camp rules plainly state that energy can be neither created nor destroyed, and since everything is made of energy, this gives us leave to relax about the resources needed to make a product, any product—including food ones. For farming, these resources refer to things like water and topsoil. The idea held by First Law folks is that resources, even natural ones, enjoy exchangeability, and if you can't swap it out, there's likely a techno-fix. Like chemicals.

Nobel Prize–winning economist and father of the neoclassical growth model (and head camper) Robert Solow said it succinctly: "The world can, in effect, get along without natural resources so exhaustion is just an event, not a catastrophe."

Of course, not everyone sees our glass of planet resources as half-full.

As lauded as Solow and his groundbreaking Presidential

Medal of Freedom–winning work is, he does have an academic foil: Professor Herman Daly, former senior economist in the Environment Department of the World Bank.

Daly sits squarely in the Second Law of Thermodynamics encampment, and its rules are predominantly tacked to the sparsely populated pup tent pole. It states that each time energy transforms—like from a seed to a plant—some of that energy is lost. Essentially, there is waste in the system. And while chemical fixes such as fertilizers may get you by, there comes a need for more and more to make up for diminishing returns, as seen before with the post–World War II farmers.

In Daly's book *Steady State*, he compared Solow's circular First Law economic models to thinking of an animal only in terms of its circulatory systems, sans output. Though it took me a couple of read-throughs, I saw what Daly was slyly telling the learned academy of neoclassical economists (a group I like to imagine wearing their velvet graduation tams and puffy-sleeved gowns). He thought their closed economic model was constipated. Full of shit.

Solow and Daly, along with their protégés, volley these erudite insults at one another via academic journals. At times, their essays read like a scholastic pissing match worthy of daytime television. But while academics battle it out from their ivory towers, farmers on the ground know firsthand that it's tough being in sustainable agriculture bound by the environmental concerns of the Second Law of Thermodynamics while functioning within an economic system entirely defined by the concerns of its First Law.

It's like a soup we all steep in. We've been acculturated to shop on price without considering its means, and not until recent history have we begun to penalize (or even arch an eyebrow toward) those who exploit our finite resources for their personal gain.

Fortunately, the message that cheap food is not cheap after all, and that natural resources have real value, is getting out there— slowly. It's a tough sell, the idea that we should spend more money now for viable growing environments in the future. If I thought a "Second Law First!" campaign would stir the nation, I would wage it, but I'm pretty sure I'd just be stuck with a lot of T-shirts.

Chapter 22

I was relieved when Dr. Stewart shifted our conversation from the theoretical to his concrete (and often-cited) research on real farmers. If our first topic explained why agriculture is broken, this second one illuminated to what extent. His data followed the nation's shift away from the early-twentieth-century model of many small, multi-crop and multi-livestock farms to the bigger, industrial monoculture model.

And it goes like this:

Say in 1910, you had a dollar to spend on food. This was back when the actual paper bill read, "One dollar in silver payable to the bearer on demand." Back then, your farmer, likely running a diverse operation, received 41 cents of that food dollar—after expenses. Another 15 cents went to farm expenses like seeds and machinery, called inputs, and then the balance, the remaining 44 cents, went to the marketing sector.

Now when I say *marketing*, banish thoughts of billboards, radio jingles, and celebrity spokesmodels. Think of it in the broader sense, as in everything that has to happen to get a raw food, say a potato, to market. The farm family might have done their own washing, quality control, and bagging, while the marketing sector was likely in charge of warehousing and getting the sacks off to retailers and ultimately to the serving dish on your dining room table.

Zoom ahead to 1990. This is postindustrialization of the food system when monoculture, specializing in raising one food for market, had partnered up with technological advances such as big machines, chemical fertilizers, and the move toward modified seeds. This was all in the name of market efficiency, as farmers could specialize on the tasks and machinery to raise one or two crops or animals, rather than tending to an entire diversified farm.

These efficiencies, it would seem, should make it easier for farmers to make a profit, right?

Revisiting that food dollar in 1990, inputs increased to 24 percent, so just about a quarter now went to fancier seeds, more chemicals, and bell-and-whistle-laden machines. And the growing marketing sector was now swallowing up 67 percent—a good-sized hunk of that buck.

Think again of those potatoes. The marketing sector was interested in the growth market of processed foods and was starting earlier in the process, picking up spuds right at the farm gate. Potatoes would catch a ride to a processing plant and start their transformation into a convenience product, such as instant mashed potatoes. Every action, from washing to peeling to dehydrating to the spraying on of sodium acid pyrophosphate and sodium bisulfite plus an emulsifier such as monoglyceride, was now handled by the marketing sector altogether. The resulting concoction would then be shaken and sealed into a labeled and logoed box with nutrition charts and put onto a series of trucks for distribution from hubs to warehouses all over the country to your store.

And how much of the cut does our 1990s farmer get? That'd be nine cents of that food buck—NINE! Put another way, if you paid the farmer nothing at all, items in the grocery store wouldn't even be a dime cheaper on the dollar.

This requires a dramatic pause.

Sadly, these numbers haven't improved for the American farmer in the past twenty-five years.

While Locally Laid hasn't been able to zoom back to those margins of 105 years ago, our partner farmers do get a bit over 20 percent of the food dollar, and that's after paying their considerable feed bill and a slim margin to us. At more than double the proportional income of most American farmers, it's not perfect, but it's a step in the right direction.

Our nation's turn toward commodity agriculture and its resultant 78 percent decrease in farmer profits led to something other than just poor farmers in this country. It also shifted land ownership. Researcher Willard Cochrane calls it the Agricultural Treadmill. He explained it like this:

As farmers crank out commodities, be it grain or eggs, there's no way to get a better price for a better product. A commodity is a commodity, after all, as interchangeable as a wing nut. Where one can game the system for more profit is to produce that wing nut on the cheap. That is, until whatever cunning technique the early adopter is using to cut costs, be it chemical or mechanical, becomes standard industry practice. Then as all the farms appropriate whatever clever idea that first farmer was leveraging for bigger yields, everyone's crop yield goes up and all are rewarded with a market flooded with the commodity and a lower price for all.

This sets forth the whole jogging process anew with the next innovation, and those farmers who can't afford to climb back onto the treadmill are, as Cochrane says, "cannibalized," bought out and folded into other, more profitable operations.

But following the dots of how this American tragedy occurred, the undoing of these midsized farms that are Middle Ag, and the gutting of communities they supported, reveals our road map out of it.

I'd called Stew back at his Maine farm but got Sarah on the phone instead. She spends her days organizing all the orders and dispatching the farm's trucks out six days a week. "Stew is very clear that the way to capture back more of the farmer's dollar is by owning more pieces of the chain. That includes distribution to stores," she said.

She also joked that the reason Stew continues to farm this way, growing a great number of crops to serve one geographic region with all its labor and headaches, "is just to prove himself right."

I hope he does.

Taking his research and using it to define what it is we've been doing with our partner farms became the crux of my presentation to the thirty or so Minnesota Cup judges at the Carlson School of Management.

After weeks of writing and rewriting our competition business plan and working with mentors on everything from public speaking to marketing, I was standing in front of these esteemed businesspeople in an intimidating lecture hall, wearing borrowed dress clothes. There I gave a sweaty twelve-minute memorized speech on how Locally Laid was actually bringing back better margins for producers.

I swept my hand across PowerPoint slides indicating how we and our partner farmers eschew vertical integration by contracting with other midsized producers working in our region. How each farm patronizes nearby feed mills and farm supply stores participating in the hippie-sounding concept of value chains, which strengthens an entire region's economy.

In truth, though it hadn't even been a year, we were already seeing material improvements in this community.

As I was giving this talk, the audience stuck with my goose-bumpy salute to the millions of Americans who continue to live rurally despite the disappearance of so many solid ag-related jobs.

Before you get concerned that we'll get bigheaded or trip over our imaginary cape, you need not worry. We came in second in the competition, solidly preserving our runner-up legacy—though it was hard to feel too bad, as we lost to a company with a nutritional drink for the gravely ill, mostly cancer patients. We also walked away with a five-thousand-dollar check, new connections, and a better way to speak about what it is we're trying to do out there in the big world.

Chapter 23

A couple of months later, I was back in Maine to dissect a chicken.

This had seemed like a good idea, on my Duluth couch, while reading the agenda for this multi-day, hands-on experience—get to know the bird inside and out. I wanted to know more. Lots more.

Now a full-time Locally Laid employee, I desired mastery with layers, to be a good steward to our girls in the field, a coach to our partner farmers, and the well-informed community resource people believed me to be. As the biggest poultry game in town, we field a lot of questions. Hobby farmers and backyard keepers call or find me out in public, leaving me to illustrate basic poultry care, like how treat a chick ill with "pasty butt" (holding her rear under a warm running faucet to wash off the fecal matter blocking her vent) all while standing in the produce aisle or parking lot. I was desperate to avoid being that Locally Laid lady who gave bad advice to the 4-H kids.

In this knowledge quest, I found resources thin. Tradition-ally, farming is an art and science that has been passed down from parents to children. But as industrialization relocated farm kids to new lives in the city, there's been a widening instruc-tional gap. First-generation farmers like us are left to library books and YouTube tutors.

I was ready for something more formal.

There were twenty or so participants at the Applied Poultry Science Project. It's a grant-funded effort to instruct New England educators, such as extension agents, agricultural program teach-ers, and mentors, in all things chicken. Clearly, Duluth, no matter how cleverly one folds the map, isn't on the East Coast, and I'm no farm agent. However, I'd found this rare three-day intensive work-shop after weeks of tickling Google.

There had been few offerings that weren't either a one-hour lecture for backyard keepers or a multimonth commercial course in another country. When I saw the Maine program, I desper-ately wanted in. After an e-mail, I followed up with a phone conversation in which mild begging may have occurred. Plus, I promised to earn my keep by teaching a marketing workshop I'd developed specifically for farmers. It's my "dog and poultry show" that I'd carted to many university classrooms and some farm and food conferences. When I offered a trade, they accepted.

But now wearing a white apron with protective glasses and booties (biohazard protection to avoid disease sharing from our own livestock), I was more anxious than excited. My stomach felt unanchored. The live birds saved out for us from a recent cull (farmer-speak for *slaughter*) looked a lot like our own chickens.

Hearty pasture breeds—no skinny leghorns here—showing a bit of road wear with unkempt feathers and faltering combs.

Michael, a poultry PhD from a few states south, led the session. He's renowned in his field and works with many farms, including large corporate operations. And he doesn't suffer bleeding-heart yahoos. The day before he'd said, "Seems like anyone who's watched a few Joel Salatin videos thinks they can start a chicken farm these days." I thought back to Jason on the couch, mesmerized by the glow of his Mac, and couldn't help but nod in agreement.

"Okay," Michael shouted above the murmur of participants, "let me show you the best way to hold a hen." He reached into the crate and grabbed one. While I instinctively hold a chicken like a football, he extended his index finger (also known as your distal phalanx, but better recognized as the barrel of your mime gun) and slid it in between the legs of the forward-facing bird. As he did, the chicken, who'd been nervously cuffing her wings, calmed.

"I'm gently squeezing her," he said, indicating her leg regions, and to further illustrate her state of relaxation, Michael turned the chicken completely upside down—and her wings flapped open without even one concerned cluck. It was like a poultry party trick and our group ooohed in appreciation.

He handed me the hen, which I lifted from underneath as demonstrated. The black-and-white-flecked Barred Rock, the zebra of the chicken world, settled in nicely. Just as I thought she felt secure and protected in my warm clutch, Michael broke my illusion.

"Chickens are prey animals, and it knows it's caught. But as long as I'm not making any moves to eat her right now, she's all right with it—right?" he said, directly addressing the bird.

While I preferred my scenario, Michael's made more sense. A captured chicken would instinctually conserve energy on what has historically been a losing battle, perhaps storing energy to bolt later. I wanted to whisper that I had no plans to eat her whatsoever, but felt disingenuous given that I was about to take an educational tour of her insides. Instead, I told her she was a pretty bird, a nice bird, as she robotically tilted her head to the soft lilt of my voice.

Michael then used my chicken for another demo. "You can take her wings up like this," he said, gently lifting them, "and cross them over one another as such." He then swirled the wings in a move that looked like the first step in a French braid, folding some of the left flight feathers over the right wing, keeping them in place. The bird, still amazingly serene, was now wings up as though sporting an exotic hairdo. This useful positioning allowed for an unfettered inspection for lice or mites, easily spied under the wing where there's the least feathering.

(Later at home, I typed "wing tie, chicken" into my Internet search engine to see if I could review a video. That was when I learned it's a term shared with the BDSM community—and was served up an entirely different movie clip.)

Passing my hen over to be placed in a rubber garbage container to be euthanized, I looked away. The trash can lid accommodated a plastic tube attached to a silver tank of carbon dioxide. It delivered a swift kick of anesthesia and, within a few minutes, death due to respiratory arrest. It was fast and seemed, from my side of the container at least, to involve little distress.

Eric, back in Minnesota, took mild umbrage with my using the term *euthanasia* this way, feeling it should be reserved for relief of great pain during a terminal illness. But looking it up, I saw it's from the Greek word *thanatos*, meaning "death," and the prefix *eu-*, meaning "good," thus giving these girls a "good death." And I can assure you that it was.

I skew squishy-hearted in many matters, but being in the company of these farm-hardened professionals held my composure firm. These were not the chicken keepers who purchase the commercially available hen leashes to take their birds on an evening stroll. No. Most had been talking nonstop about hunting or processing animals or culling flocks since we'd met yesterday, like it was as natural as wearing camo to the feed store.

As we entered the hay-scented cow barn, I headed for the pre-set lab table that the Heidis had claimed. The Heidis were a pair of affable and funny gals who'd driven up for the conference from Rhode Island. When we did our group introductions and were instructed to "add something memorable," one Heidi shared that she'd once been both the fair queen and tobacco-spitting champion at her county fair—the same year. It would be foolish to even try not to like them. And now, more than ever, I needed to be distracted from the dizzyingly large and sharp shears on the tables.

Handed a dead chicken, this time a cinnamon-colored bird who looked a bit scraggly, I silently offered her thanks for her service and looked up to see that most of the other participants had already dug in. I exhaled and got to work.

The dissection was narrated by Michael, who talked us through the first resistant crunches of the rib cage. I concentrated on his voice to avoid focusing on the grisly sounds of breaking bone. But once the bird's impressive chest cavity was revealed, I was taken with the chicken's tidily packed organs. Snipping out the lungs and heart, I'd gently set them aside, in order, like I was dissembling a broken toy and I might be asked to reassemble and perhaps even reanimate her later.

With Michael's help, I located the crop, the store for food before it goes onward for digestion. It was a yellowish sack protruding from the esophagus, and I heard someone at another table refer to it as the bird's built-in to-go bag. Opening it up, I scooped out a sloppy clump of wet grains, her day's meal, as other tables pulled out amazingly long strings of grass.

That's a risk for free-ranging chickens.

They pluck and swallow blades of grass, which can lead to a condition called crop impaction. That's when food fails to pass through on its digestive carnival ride through the body. It can be due to a few things, but for outside birds, it's often that the organ's exit is blocked by something like wood chips, twine, or stringy plants. It's a condition that can take a healthy chicken and waste her to nothing. To prevent this at our Mom, Pop, and Bro operation, we pass a mower over an overgrown pasture, keeping the field to no more than six inches before bringing in the flock. This is in lieu of cows or some other grass-chewing ruminants. I'm pretty sure that they're in my future, but I pushed this out of my mind, too. I needed to concentrate.

I continued on down our bird with the other woman at our

table, Nicole, an organic certifier from Vermont. She was a capable and curious learner, and I tried to vampire off her eager presence. Hitting the lower abdomen area, we waded our way through large and copious globs of yellow fat. Nicole shouted out a question: how can outdoor birds, presumably exercising every day, accumulate so much fatty tissue?

"Because they're bums!" piped up Dr. Michael from across the barn. "Hanging around eating all day and not giving eggs."

In the poultry world there are two kinds of mature hens: layers and loafers. Earlier, Michael had flipped over a live bird and showed us several indicators of infertility such as a dry vent opening and stiff pubic bones that would not allow an egg to pass. Then he swung her back up to examine her comb. A bright red headdress indicates fertility, while a light-colored, scruffy one signals that she's spent. The fading mane and disposition toward heaviness around the middle and backside are parallels not lost on this middle-aged woman.

Continuing with the dissection, we pushed on to the gizzard, a tube Nicole neatly sliced wide with our cutting shears. We were to look for worms. After poking around with a gloved finger, we declared our bird clean. Michael was impressed that only about a third of the flock hosted these wiggly moochers. The Heidis showed us a sample from their chicken's gut—three or four thin, long, white strings, a small infestation given that worms work to entirely encircle intestines until a well-eating chicken starves.

"If you don't want your chickens to miss out on having parasites, put them outside," said Dr. Michael with a little snark. I bristled under the pointed criticism but couldn't deny his point. While chickens outside on pasture encapsulate the beauty that is poultry in motion, it's sadly also the perfect medium for free-loaders like lice, worms, and mites. It's safe to say that nearly all outdoor flocks are fighting some sort of bug.

As chickens first started making their way indoors and into confinement, they exchanged the lively feeling of wind in their wings for actual life. Mortality rates in outdoor laying hens dropped from 40 percent in the early 1900s to a mere 5 percent in modern caged operations. And that's a number that's hard to argue with. Especially with feisty Dr. Michael.

But would that chicken choose to have every indoor need met or face a more scrappy existence outside? Some of our birds have to be coaxed out of their shelters, perhaps avoiding interaction with bigger birds or reluctant to give up a particular sweet spot on a roost or some poultry scenario simply unknowable to me, like a bad feather day. But I can tell you that without fail, when our barn doors open in the morning, there's a head-bobbing, goofy stampede of fowl—even into inclement weather. One could argue that in a chicken's walnut-sized brain there's no room for critical thinking regarding bacterial microbes and parasitic risks. I wouldn't argue that. Though her instincts tell her that there are hawks and foxes out there and despite the ample feed, water, and safety in the barn, she still ventures out. Chickens, no doubt, have pluck. Though I would think on Michael's words for months to come.

There were other standout moments at the conference,

several with another poultry scientist, Paul. Tall and lean with salt-and-pepper hair, Paul had an easy manner while walking us through his slides of free-range pasture projects he had been doing in Pennsylvania.

"My grad students are supposed to move these chicken tractors around the field," he said, showing a photo of a portable henhouse. "But somehow it always ends up being me." He laughed easily at himself.

"Our findings show that eggs from these outside birds had more omega-3, vitamin A, and vitamin E."

And before my smile was fully formed, he interjected, "But, you know, I could just add that to the feed."

I sighed. He's right, of course. There's a national egg company that has made its name on pumping its eggs full of chemical vitamins and minerals, even adding marigold to give its yolks the orangey hue that pasture-raised chickens earn outside.

However, I have no doubt that eggs from hens outside truly are better. This sprinkling in of vitamins feels akin to the story of baby formula. When I was an infant in the 1970s, I was bottle fed because the collective consciousness embraced a "better living through chemistry" ethos. Recently I noticed that formula cans are now splashed with claims of new additives like DHA and ARA fatty acids for brain and eye development, which might explain my eyewear and inability to juggle numbers in my head. One would guess these acids have always naturally occurred in breast milk, but it took some forty years for scientists to replicate them. So, when it comes to pasture-raised eggs, other than those fatty acids and vitamins that present-day scientists are all about, what other beneficial properties are going on in the yolky down-low that we can't yet test for, much less make in a lab?

My most startling piece of incidental learning happened later that day in our hotel conference room. A local food program director and small-scale pasture bird farmer asked a specific question about a new flock of chickens. She described a familiar scene—piling into corners, aggressiveness, and all and all being a difficult, nervous flock. This was unusual, she said, as they'd had several cycles of birds and had always enjoyed mellow hens.

I leaned forward. For one, I rarely met farmers who raise chickens the way we do and I was interested in her difficult flock. But I was even more attentive to the ideas that these troubles were an *exception*, not the rule.

"Sounds like they were not properly lighted as pullets," said Paul.

An entire back-and-forth ensued that I watched like a bug-eyed Ping-Pong tournament fanatic. It focused on the light-tight barns where commercial chicks are typically raised. The artificial light showered on pullets is an entire science, complete with charts and graphs and interactive computer applications that will pinpoint a brooder house's position on the earth for exact sunrise and sunset information. It's that important. While I understood that light affects a bird's pituitary gland and the release of her reproductive growth hormones, what I didn't get before is how important it is to adjust for birds destined to be outdoors. Otherwise, a hen will go from a hormone trickle to a virtual estrogen tempest, which sends their little poultry nerves into a hysterical swirly.

Poultry hysteria is a terrible condition that entails everything our birds suffered through in those early days on pasture. It cranks up perfectly lovely chickens to acts of cannibalization, piling and suffocating in droves, and becoming flustery, wing-batting freakouts to every actual or imagined sound or movement.

"Good Lord," I said involuntarily.

When the next break was called, I bolted out of the room. Holding my aging cell phone in front of me, I paced the circular driveway of the inn, desperate for a signal. When I found two bars' worth, I dialed Jason.

"Hey, Bird," he answered. "How's Maine?"

"It wasn't our fault." I talked quickly, with the realization that I had far more to say than the cerebral bandwidth to say it.

"What? I don't think we have a good connection." I heard our own Minnesota chickens clucking in the background.

"The pullets—all their problems—they were improperly brooded and I know I'm not being clear here, but Jason, it wasn't our fault. The smothering, the blowout, it wasn't space issues or diet."

"Myron's birds?" he asked, catching up in the way only the long-married can.

"Yeah." I nodded into the phone. Surprised by the bubble of emotion that forced its way up my throat, I pulled my red jacket tight around me and said, "Yeah, you know, if we brooded our own, we could do it right."

"We could do that when we live at the farm," he said.

Yeah, about that.

I'd tried to picture it: our family out in rural Wrenshall living the happy ending that everyone wanted. The solid home, the sweeping landscape, all among our beloved birds. Over the past weeks, as I wanly smiled over these bucolic images in my head, there'd been something worrisome around the edges.

Then it became clear. Maybe it was hanging out with my new conference friends, people who took such joy in their agricultural pastimes and simply infused the air with their capability that allowed me to see it more clearly. I truly admired them. I thoroughly enjoyed their company. But while I wasn't exactly a fraud, I wasn't exactly a farmer, either. I left Maine remembering something important. I never wanted to move our family out to the country.

"Jason," I said, cornering him in the kitchen when I got home. Thus began a carefully worded, tender speech outlining our life events, starting with my giving up a Twin Cities writing career and moving to Duluth—for him; then starting this farm with all its travails—again, not my life goal. And while I admitted that these things had turned out more or less okay, if you don't count the years it sheared off my life, there had also emerged an undeniable theme.

"I know you think it's another case of my resisting change or maybe not being brave, but actually, I've been thinking about this for weeks now and"—I swallowed—"and, Jason, moving to Wrenshall is simply not my dream."

His response was physical. I could see a happy, hopeful

something in his body drop low, pulling the confidence in his chest with it. I had to hold myself back from chasing after this man I loved as he stomped downstairs into his office. But at that moment, I held another someone dear to me in the forefront of my mind: Abbie.

Although she was up for the move, I didn't want her to watch her mother, her closest female role model for how a couple negotiates a marriage, take her life in a direction she didn't want—again. I'd already walked right up to the edge of where compromise meets pushover and Abbie, unknowingly, was the ballast keeping it all from tumbling over.

The next days, I endured a household anxious under a communication blackout. Conversation was limited to the necessary subjects of kid pickup and chicken chores. It remained that way until Jason came to me as I made dinner.

"I've been thinking," he said, slouched into the corner of the breakfast nook, "and you've done enough, probably too much."

That turned my head.

"So, Bird, if you want us to keep living in Duluth, then we should."

There is nothing more handsome than a man who thinks I'm right. But what put the lump in my throat was his acknowledgment that I had given, had indeed sacrificed. I rushed him, pushing him against the counter with a hard kiss.

But that still left the question of the farmhouse. Not every tenant wants to share a yard with a couple thousand chickens.

Then just as Jason had once saved Brian in Cambodia and

Brian, in turn, saved Jason on the pasture that first winter, Brian quietly saved me.

He approached us about moving his young family, now with a second baby daughter, Daisy, out to the farmhouse. A few months later, they did.

Chapter 24

On the cusp of winter in November 2014, Jason came home dirty from farm chores and wearing a sour face.

"I just got a complaint call about the eggs," he said. "The chef told me they're just not what they used to be."

This stung, mostly because I knew it to be true. While I wasn't washing eggs every week, the last time I subbed in I'd noticed a weakness in the shell—and recognized it for what it was: a symptom of advanced poultry age.

"It's probably time. It was time a while ago," I said.

Eggs are a young bird's game. Some of our chickens, the elder statesgals of the flock, were now twenty-eight months old, and their advanced age showed with less production. And what was left in the nesting boxes was of poorer quality. We'd kept them nearly a year and half beyond industry standard. The chickens were now geriatric layers.

He nodded.

We revisited all our options for culling the flock and reminded ourselves that there were several reasons to do this. Beyond the obvious egg problems of an aged flock, it was also time to thoroughly clean our barn and let it sit vacant for a bit. This is the best way to eliminate parasites and tamp down disease.

While the barn had been a huge improvement over wintering in the field, it hadn't been built to house hens and thus wasn't properly insulated. We hadn't finished paying the last winter's propane bill until August. With a vacant barn, we could reevaluate our options.

But first, we had to empty that barn.

While we liked the idea of bringing our chickens to the USDA-certified facility to be processed into stew hens, that would mean renting a truck and bird crates and driving them at night (for the least amount of poultry stress) a few hours west. Then spend a night in a hotel and bring them back in a series of coolers—which we'd have to buy. Or renting a second truck, this one refrigerated. Suddenly, we were looking at expenses in the thousands for a product we weren't sure we could even give away.

We reached out to our contacts in the Asian community. Although it took us a while to find the right person, he agreed to come take our hens live off the farm and pay us a modest buck a bird. They would use these pasture-raised chickens (tough as they were) to feed themselves, and this felt, if not exactly good, then fitting.

I liked the idea of providing high-quality, low-cost protein to a community willing to do the work of processing it. We fed the chickens another two weeks as we attempted to confirm a date. When the set day finally arrived, he didn't show.

"Brian talked to the dog guy," Jason said. I looked away. Giving our girls over to be dog food for a musher wasn't my first choice. It didn't even make my top five. But it felt better than the advice we'd gotten from other farmers: kill them and lay them on your fields as fertilizer. There was some sense in this, as we were looking to diversify and start the process of adding berry plants to our farm. But it seemed wrong to halt a good bird's "circle of life" train short and with it the opportunity to pass on her superior nutrition before going back to dust.

The next day, the dog man came and Jason returned home quiet. We both were, but we also understood this was farming.

It was a desolate period being chickenless chicken farmers (as our partner farmers filled the gap while we were out of lay), and that was when I received the envelope. It came to our Wrenshall post office box addressed to "Management," a demarcation that made me smile. There's considerable overlap between labor and administration at Locally Laid Egg Company.

I opened up the envelope and found a handwritten letter.

Dear Locally Laid,

I find your name on your egg carton extremely offensive and your sexual innuendo in advertising them vulgar. Not only were they the highest price in the store but also the worst in advertising. I will share this message with the owners of the

grocery store and friends. We have enough crudeness in the
world without egg advertising adding to it. I bought them in
a hurry and will not again.

Most people really liked our Locally Laid name and it's fair to
say we received several hundred compliments and energetic
thumbs up, easily balancing out the three complaints we'd gotten over a couple of years. But this one felt different. Maybe it
was the steaming vitriol behind the angry penmanship that
made this gentleman feel more real to me. Like he was an actual
person: one I'd hurt.

Having lived in the Midwest nearly twenty years, I'd finally
picked up the complicated steps to its social dance, though not
without a few stumbles along the way. Initially, the blunt New
Englander in me was completely tone deaf to the subtle dodging
of conflict, always skirting what one really means and never
stepping on anyone's toes. Ever.

I knew the contrite response I was supposed to write.

Even so, I'd come to think of Locally Laid as "dirty optional."
I mean, you can go there if you want, but there's also a perfectly family-friendly primary meaning that thoroughly describes
our venture. But while I'd meant to push the limits of social
acceptability, I hadn't intended to brutishly burst through them,
either.

I decided to sit on it. In fact, it was a full month until the
guilt over a timely reply got to me. I'd set up the kids with beanbag chairs and popcorn for the president's State of the Union
address and then dug out the letter.

As the official procession fanfare started in the chamber of the House of Representatives, I reread the rant and found myself having an entirely different set of reactions. While I still didn't like the idea of vexing people, even prudish ones, I was stuck on one phrase, "the highest price in the store."

You know, I'll take my lumps on the double entendre, justified or not, but when it comes to the cost there are many good and real factors that go into that number. Yet our eggs still come in pricewise well below many other cartons on the dairy shelf. After I'd crafted a sincere acknowledgment that my letter writer had every right to stand up for his opinions and indeed more of us should, the televised speech got more rousing. And that was when I started typing faster.

Milo turned to shush me, so I moved into the dining room, where I could still hear the inspirational cheers from our nation's capital. Fueled by the rousing applause, I deconstructed our moniker starting with "local" and explained all I'd learned about food miles. I also informed him of our planting of what was then some three thousand trees to offset our minimal carbon footprint—before ever taking a paycheck.

I went on to explain how sustainable agriculture largely misses out on large government subsidies that commodities enjoy. And the difficulty and risk of breaking into this highly consolidated industry, one that by its nature of keeping chickens by the hundreds of thousands can more easily spread a disease like salmonella over millions of eggs, sold under multiple labels, resulting in a large-scale food safety nightmare.

Then I moved to the labor, the fence moving and the feed

buckets. Also how our small flock sizes benefit both our poultry athletes and their eggs, but these same practices pit us against factory farms that enjoy massive economy of scale.

By this time, I really had my Middle Ag on—it felt like LoLa, that inspirational logo chicken, lit on my shoulder, bolder than me, drove my typing. From my keyboard, I clacked our information about shuttered midsize producers and the rippling effect on rural communities, from feed stores to schools to corner cafes. Moving on, I hit on the importance of sourcing local inputs and selling locally, business to business, keeping all that money swirling in small communities rather than extracted from them. And the financial risk we took on as we worked to create shelf space for our partner farms—and fetch a fair price for their goods.

Most importantly, I admitted how our farm isn't perfect. That we wished there were more we could do for our wintering birds than just providing roosts and dust baths and bringing dried prairie grass into the barn. I wistfully yearned to entertain our chickens by teaching them games like Toss Across.

In closing, I wrote:

> *So, to the point of your letter, I want to say you're right. Our name, Locally Laid, is totally cheeky and pushes the envelope. And I truly am sorry we offended you. (I'd offer you one of our American-made Local Chicks Are Better T-shirts, but I don't think you'd wear it.)*
>
> *But here's why we risk your umbrage. When our perfect double entendre breaks through the media clutter in which*

we're all steeped, we leverage it. With that second look from a consumer, we educate about animal welfare, eating local, real food and the economics of our broken food system.

We all vote with our food dollars every day and we respect your decision if our playful moniker keeps you from buying our eggs. It was just important to me that you understood everything that was going on behind that name.

Now I gotta ask, would you have learned all this if we were named Amundsen Farms?

Jason read it and said, "Wow, you really went full-frontal farmer on this guy." He then put a hand on my shoulder. "I know you're going to make this into a blog, but I don't want you to be disappointed. It's unlikely a lot of people will care at this level."

Yeah, at the heart of it, this was a pretty wonky response and when I threw it online I figured about fifteen folks, mostly other farmers I knew, would read it all the way to the end.

Then we left for the bank. We needed to increase our line of credit, not because we weren't selling enough eggs to cover expenses, but to bridge the gap until we got paid for those eggs.

To that end, we'd romanticized Locally Laid's financial system as thoroughly as we had our pasture one. We'd envisioned a graphic-worthy arrangement that shuttled our superior product to retail and wholesale outlets, and payment arriving in a timely manner allowing us to stay square on all our bills. It didn't work that way. Sometimes stores would wait over forty-five days to pay us, completely ignoring the fourteen-day terms we'd specified. Or they would cut the check on time, but not mail it for weeks so their books looked like they were caught up on bills.

The financial stress was crushing us.

As I sat in that familiar bank chair, my phone started pinging, indicating every Twitter share and Facebook comment. Jason eyed me and I turned my phone off with an apology to our banker.

By the time we were walking out twenty minutes later, the blog had been shared hundreds of times. Just three days later, social media numbers indicated views of over 250,000.

Jason and I giggled like children, even as the response crashed our website.

Soon, the media were calling and I was sitting in the Duluth branch of our public radio station for a statewide interview. I was tired, having been up well past midnight stuffing T-shirts into envelopes, fulfilling orders we were getting from all over the country.

I was looking forward to hanging out in the studio with my friend Dan, the radio reporter, but he'd left to work on a story. There was an unreassuring note on top of the large panel of buttons and levers that read: *Don't touch anything!*

As I sat gingerly on the chair in front of the large mic, the producer 150 miles away rang my cell and asked, "So, is it 'agriculture of the middle' or 'middle agriculture'?"

"We're going to talk about that?" I asked.

"Yeah! We're really excited about it," she said from our state's public radio mother ship in St. Paul. "I wouldn't be surprised if the segment goes long, so be prepared to talk for up to twenty minutes."

I gasped a little.

While I was more comfortable with public speaking from the last couple of years, most of these media interactions were two minutes, three tops. Twenty minutes felt like an eternity.

"Okay!" I said, hanging up as my stomach banked a corner without me. I regretted having eaten breakfast.

It clipped along quickly and went just fine, even after that unexpected and frightening sentence: "We'll open the line to callers."

After a few media cycles involving TV, podcasts, and blogger interviews, we received incredible feedback. Egg sales spiked and we sold some five hundred tees in thirty-seven states. It meant we didn't have to dip into that new line of credit. In fact, we even took our first small payday some two years and five months after selling our first egg.

That night, we took the kids out for dinner.

"So what's our hourly rate work out to be?" I said, indicating the check Jason was showing the children.

He smiled a "don't you even start" smile and ordered us a pitcher. We were, after all, celebrating.

A few times a week, people stop and congratulate us on our "success," mistaking our poultry fame for big farm profits. We're still a very new company, after all, with significant debt to service and investments to make. And though I joke with them that we make tens of dollars from Locally Laid (tens!), I know they don't believe me. They don't want to. It's not the ending they want and, frankly, I wouldn't, either. But surely we're not yet at the end. Profits will come, just not today. But I don't push this onto these kind people, legitimately excited for the farm. I let them think what they want—that I'm modestly downplaying our financials.

Occasionally, someone will take it further and tell us how lucky we are to have started the farm at all. I smile and usually say something along the lines of "It's been a mixed bag," hinting at its strain. Of course, if I were to rummage through that diverse sack of experiences, I'd fish out good feelings, ones of doing something important, something physical, something bigger than just making a profit, or even having our bit of "poultry celebrity," which I quantify as being about four clicks below the lowest-rated weatherman on the public access channel.

However, it wouldn't take long to run into something cutting within that mixed bag representing the physical hardship, as well as the financial and relationship peril. I'm glad we can't go back in time, because I don't think we would do it again. When Jason jokes that the farm was his midlife crisis and it was either that or have an affair, I find myself wondering which would have been more difficult on our family.

Because saying, "It's been hard," well, it's not sufficient. The stressors have felt real enough to have their own chair at the dinner table. But mostly, I missed my family during the startup, just wanting us to be physically and emotionally together, like before the farm, with time and money for milestone trips to

Yellowstone and Washington, D.C., much less a simply frivolous excursion to the Harry Potter theme park.

Mercifully, though, our day-to-day lives have gotten physically easier in the past year. Abbie will start high school in the fall of 2015 as Milo continues to plug away at his middle school years. Perhaps after spending the summer wrangling new LoLas and learning about crops, they'll welcome the relative ease of classrooms and book bags. There's been talk of their setting up a farm stand and even hope that our eldest may apply for a farm work license, allowing her to drive sooner than her peers. Mom says, "We'll see."

Locally Laid is even in the process of hiring some full-time positions, adding a couple of living-wage jobs to our growing Northland economy. This should help greatly with the workload, but I suspect we'll never completely relax, always an ear cocked for the other work boot to drop—an out-of-order Aqua-Magic, a fouled-up invoice, or problems to work through over a wide cultural divide with our Amish partners.

Thankfully, I'm not alone for any of it.

Jason is with me, still full of wild ideas and farm-enhancing visions—Sustainable berries! Garlic! New chicken feed formulas!—but most importantly to whisper in my ear the encouraging truth: everything's coming together.

Indeed, it's right around the corner.

Acknowledgments

Writing a book takes a clown car's worth of supporters to hurl it forward. I'm grateful to those who crammed in for the ride.

Right out of the gate, I want to thank my parents for my liberal arts education. It made all the difference.

Sincere gratitude is also expressed to:

My editors, Caroline Sutton and Brianna Flaherty, and all the fun folks at Avery/Penguin who guided this project and let me keep Jason's mugging scene.

My agent, Holly Bemiss, who made me rewrite my proposal more times than either of us care to remember. You were right.

Gail Blum, John Erickson, and Rod Graf, for entangling yourselves in this farm project and our lives. Without you, there would be no Locally Laid.

Also all our volunteers who built our hoop coops for only lunch in a barn and a T-shirt.

Patricia Weaver Francisco, my thesis advisor and friend, who greatly influenced this book, along with the entire creative writing program at Hamline University, especially Dean Mary Rockcastle, who awarded me the Richard P. Bailey scholarship the semester I truly needed it.

Dean Bill Payne and the University of Minnesota Duluth for the flexible work schedule to write.

The Glensheen folks, who endured a chronically underslept co-worker and never made me close the mansion alone in deference to my bravery impairment.

Dr. Randy Hanson for beginning my middle-agriculture adventure.

Eric Ringham for encouraging me to tell the big story of farming through my little story of farming, and for putting my essays on Minnesota Public Radio.

My many early readers of messy drafts, but especially the keen-eyed Michael Creger, who wouldn't let me get away with a dang thing, and Tom Schierholz, who misses nothing. The book is better for it.

The Arrowhead Regional Arts Council for supporting this project with the McKnight Foundation and the Minnesota State Legislature's general and arts and cultural heritage funds fellowships.

The University of Maine's Cooperative Extension.

Dr. Stewart Smith and Professor Sarah Redfield of Lakeside Family Farm for graciously talking economic theory during high harvest.

My encouraging Facebook universe: co-workers for those who work alone.

Intuit and their Small Business Big Game competition that afforded us remarkable opportunities, as did the Carlson School of Management's Minnesota Cup contest.

The beautiful city of Duluth by the Unsalted Sea and every person who rolled down their car window at a red light to tell me they'd voted for us.

Tom Hanson at the Duluth Grill and the fine folks at the Whole Foods Co-op in Duluth for being those crucial early adopters.

Matt Olin for lending his talents to create our LoLa, a chicken who became so much more.

My mother-in-law, Jean McCue, for her nonstop encouragement.

Brian for breaking his back caring for LoLa, trusting me with his story, and the Willie Nelson joke.

My children, who endured an unreasonable number of mismatched socks while their mother was rolled over her Mac like a fiddlehead. Thank you for your good humor and for letting me write about you.

But especially, I am appreciative of my adventuresome husband, Jason. Life with you is never dull. Let's try it, shall we?

Notes

CHAPTER 1

95 percent of America's chickens are in battery cages: Kathy Stevens, "Cruelty Is Cruelty, Any Way You Slice It," *Huffington Post*, October 8, 2014, accessed April 15, 2015, http://www.huffingtonpost.com/kathy-stevens/chickens-cows-cats-and-do_b_5929814.html.

pastured hens foraging on fresh grasses producing healthier, delicious eggs with less fat and cholesterol: "Research Shows Eggs from Pastured Chickens May Be More Nutritious," Penn State University, accessed April 15, 2015, http://news.psu.edu/story/166143/2010/07/20/research-shows-eggs-pastured-chickens-may-be-more-nutritious. "Pastured Poultry Products," Sustainable Agriculture Research and Education, 1999, accessed January 2, 2010, http://mysare.sare.org/mySARE/ProjectReport.aspx?do=viewRept&pn=FNE99-248&y=1999&t=1.

CHAPTER 3

the three-year rigor to become certified: "National Organic Standards," USDA Agricultural Marketing Service, accessed April 15,

2015, http://www.ams.usda.gov/AMSv1.0/ams.fetchTemplateData.do?
template=TemplateN&navID=NOSBlinkNOSBCommittees&right-
Nav1=NOSBlinkNOSBCommittees&topNav=&leftNav=&page=
NOPOrganicStandards&resultType=&acct=nopgeninfo.

**only about 0.7 percent of the roughly two million farms in the
United States are certified organic:** "2012 Census Drilldown:
Organic and Local Food," National Sustainable Agriculture Coalition,
May 16, 2014, accessed April 10, 2015, http://sustainableagriculture
.net/blog/2012-census-organic-local/.

**most of the organic grains used to feed animals in this country
are—brace yourselves—from China. India, too:** Dan Charles,
"Chickens That Lay Organic Eggs Eat Imported Food, and It's Pricey,"
NPR, February 27, 2014, accessed February 27, 2014, http://www.npr
.org/sections/thesalt/2014/02/26/283112526/chickens-laying
-organic-eggs-eat-imported-food-and-its-pricey.

**we import as much as eight times as many organic grains as we
grow:** "Organic Feed-grain Markets: Considerations for Potential Virginia Producers," Virginia Cooperative Extension, accessed April 27,
2015. https://pubs.ext.vt.edu/448/448-520/448-520.html.

"Local supports a great many more values than organic": "Book
World Live," *Washington Post*, accessed May 12, 2015, http://www
.washingtonpost.com/wp-dyn/content/discussion/2006/04/07
/DI2006040700562.html.

**taxpaying people manage to fund food additives like high-fructose
corn syrup, corn starch, and soy oils at the rate of $17 billion a
year:** "Billions in Farm Subsidies Underwrite Junk Food, Study Finds,"
Huffington Post, accessed April 27, 2015, http://www.huffingtonpost
.com/2011/09/22/farm-subsidies-junk-food_n_975711.html.

CHAPTER 4

**Farmland values have steadily increased over recent years, some
748 percent since 1987 to 2014:** "Farmland Prices Deflating: First
Decline in Three Decades," Breitbart, accessed May 12, 2015, http://

www.breitbart.com/big-government/2015/02/25/farmland-prices
-deflating-first-decline-in-three-decades/.

Others have published books pointing out the famous couple's advantages along their "Good Life" road: "The Truth behind the Backwoods 'Good Life,'" *Bangor Daily News*, accessed April 26, 2015, http://archive.bangordailynews.com/2003/11/17/the-truth -behind-the-backwoods-good-life/.

a hen in lay will produce an egg every twenty-six hours: "Raising Chickens for Egg Production," Extension, accessed April 26, 2015, http://www.extension.org/pages/71004/raising-chickens-for-egg -production#.VTk_0WTBzGc.

research published in the Northwest Monthly . . . changed everything: "Northwest Station Produces Record Hen," *Northwestern Monthly UM Crookston*, accessed April 26, 2015, http://nwmonthly.umcrookston .edu/Northwest Monthly 1928 Vol 12 No 12 November.pdf.

Chickens need vitamin D to absorb calcium: "Poultry Health and Disease Fact Sheet," Government of Saskatchewan, accessed April 26, 2015, http://www.agriculture.gov.sk.ca/Poultry_Health_Disease.

Milton H. Arndt, Illinois farm boy turned New Jersey inventor: "Lone Girl Raises 15,000 Chickens in Indoor Cages," *Modern Mechanix*, January 1, 1937.

Present-day cages often house five to ten hens in a wire crate often 2.25 feet by 2.25 feet and 14 inches tall: "Interstate Egg Fight Erupts over Cramped Hen Cages," Pew Charitable Trusts, accessed April 26, 2015, http://www.pewtrusts.org/en/research-and-analysis /blogs/stateline/2014/11/04/interstate-egg-fight-erupts-over -cramped-hen-cages.

268 companies accounting for 95 percent of the nation's 305 million laying birds: "Welcome to the American Egg Board—Industry Overview," American Egg Board, accessed April 26, 2015, http://www.aeb .org/farmers-and-marketers/industry-overview.

2,500 egg producers around in 1897: "Eggs Profile," Agricultural Marketing Resource Center, accessed April 27, 2015, http://www.agmrc .org/commodities_products/livestock/poultry/eggs-profile/.

CHAPTER 6

Raised naturally, chickens will molt all on their own, losing and growing fresh feathers during the fall as days shorten and temperatures drop: Gail Damerow, *Storey's Guide to Raising Chickens: Care, Feeding, Facilities* (North Adams, MA: Storey, 2010), 129.

the chickens they see at the supermarket riding the rotisserie carousel are usually just a few weeks old: Damerow, *Storey's Guide to Raising Chickens,* 73.

given all the health problems these chickens have from their rapid weight gain: "The Cornish Cross: What Is Wrong with This Picture?!" *Modern Homestead,* accessed April 26, 2015, http://www.themo dernhomestead.us/article/cornish-cross.html.

CHAPTER 7

it was that same sense of self-preservation that would tell her to seek height: "Caring for Chickens: Are You Curious about What It Takes," Raising Chickens, accessed April 26, 2015, http://www.raising -chickens.org/caring-for-chickens.html.

CHAPTER 8

Less than a hundred years ago, small to midsized food producers numbered nearly six million across America: Roberto A. Ferdman, "The Decline of the Small American Family Farm in One Chart," *Washington Post,* September 16, 2014, accessed January 12, 2015, http:// www.washingtonpost.com/blogs/wonkblog/wp/2014/09/16/the -decline-of-the-small-american-family-farm-in-one-chart/.

our nation's entire food system pivoted on World War II: Andrew Kimball, *The Fatal Harvest Reader: The Tragedy of Industrial Agriculture* (Washington, DC: Island Press, 2002), 125.

Ford-Ferguson saw an opening in the market and began producing a line of easier-to-handle tractors: "Outtakes Ford Ferguson—Massey Ferguson," *Legacy Quarterly*, accessed April 27, 2015, http://www .legacyquarterly.com/LQ/Outtakes-Ford-Ferguson.

Our Gothic farm couple at this time would have had money as wartime farm incomes nearly tripled: Bill Ganzel, "Farmers Produce More Food for War in World War II," *Living History*, accessed April 27, 2015. http://www.livinghistoryfarm.org/farminginthe40s /money_02.html.

They needed them to replace the many hands not returning to the farm, opting instead to pick up pencils the GI Bill bought for them: "Education and Training," U.S. Department of Veterans Affairs, November 1, 2013, accessed April 27, 2015, http://www.benefits .va.gov/gibill/history.asp.

DDT and nitrate fertilizers, now made in former ordnance factories, were promoted as yield-boosting, labor-saving options: Kimball, *The Fatal Harvest Reader*, 95.

The Marshall Plan . . . had the United States buying billions of dollars' worth of produce to ship to rebuilding Europe: Bill Ganzel, "Agriculture Supplies Material for the Marshall Plan," *Living History*, accessed April 27, 2015, http://www.livinghistoryfarm.org /farminginthe40s/money_07.html.

Increasing amounts of nitrogen were being tilled into fields, with fewer returns: Kimball, *The Fatal Harvest Reader*, 95.

What wasn't wielded internationally was sold cheaply to domestic food processors, becoming the ubiquitous ingredient high-fructose corn syrup: Shea Dean, "Children of the Corn Syrup," *The Believer*, October 1, 2003, accessed April 27, 2015, http://www.believ ermag.com/issues/200310/?read=article_dean.

With those protections gone, farmers were exposed to the caprice of the marketplace: Tom Philpott, "A Reflection on the Lasting Legacy of 1970s USDA Secretary Earl Butz," Grist, February 7, 2008, accessed April 27, 2015, http://grist.org/article/the-butz-stops-here/.

1987 became the year that the U.S. Department of Agriculture (USDA) recorded the highest number of farm bankruptcies in its history: Jerome Stam and Bruce Dixon, "Farmer Bankruptcies and Farm Exits in the United States, 1899–2002," U.S. Department of Agriculture, January 1, 2004, accessed April 27, 2015, http://ers .usda.gov/media/479214/aib788_1_.pdf.

CHAPTER 9

egg-bound hens: Kathy Shea Mormino, "Chicken Egg Binding: Causes, Symptoms, Treatment, Prevention," Chicken Chick, July 20, 2012, accessed April 27, 2015, http://www.the-chicken-chick.com/2012/07 /chicken-egg-binding-causes-symptoms.html.

Prolapse is when a hen's oviduct is pushed inside out and protrudes from its vent: Gail Damerow, *The Chicken Health Handbook* (Pownal, VT: Storey, 1994), 52.

Prolapse was a problem in underweight birds: "Common Laying Hen Disorders: Prolapse in Laying Hens," Alberta Agriculture Food and Rural Development, accessed April 27, 2015, http://www.agric.gov .ab.ca/livestock/poultry/prolapse.html.

most things we pick up from the grocery store travel between 1,500 and 2,500 miles: "Globetrotting Food Will Travel Farther Than Ever This Thanksgiving," Worldwatch Institute, accessed April 27, 2015, http:// www.worldwatch.org/globetrotting-food-will-travel -farther-ever-thanksgiving.

CHAPTER 11

Hens have sperm host glands: Julie Gauthier and Rob Ludlow, *Chicken Health for Dummies* (Hoboken, NJ: Dummies, 2013), accessed April 26, 2015, http://www.dummies.com/how-to/content/starting-with-the- chicken-and-then-the-egg-growth-.html.

bakers liked them for the tall meringue: Rick Nelson, "A Miraculous Meringue," *Star Tribune,* September 1, 2013, accessed September 1, 2013, http://www.startribune.com/a-miraculous-meringue/222529231/.

CHAPTER 13

the reason why EU eggs aren't washed at all: Nadia Arumugam, "Why American Eggs Would Be Illegal in a British Supermarket, and Vice Versa," *Forbes,* October 25, 2012, accessed April 26, 2015, http://www.forbes.com/sites/nadiaarumugam/2012/10/25/why-american-eggs-would-be-illegal-in-a-british-supermarket-and-vice-versa/.

After starting their vaccination program in 2009, the Brits dropped their infection rate to only 1 percent in their flocks: "Half of Egg-laying Hens in U.S. Have Not Received Low-cost Salmonella Vaccine," Associated Press, August 25, 2010, accessed April 26, 2015, http://www.cleveland.com/nation/index.ssf/2010/08/half_of_egg-laying_hens_in_us.html.

CHAPTER 15

side-by-side study between conventional and organic crops conducted over thirty years by the Rodale Institute: "Farming Systems Trial," Rodale Institute, accessed April 27, 2015, http://rodaleinstitute.org/category/rodale-institute-projects/farming-systems-trial/.

we've long produced enough calories to feed the world: Mark Bittman, "How to Feed the World," *New York Times,* October 14, 2013, accessed April 2, 2015, http://www.nytimes.com/2013/10/15/opinion/how-to-feed-the-world.html?pagewanted=all.

we've more than doubled our food waste: Roberto Ferdman, "Americans Throw Out More Food Than Plastic, Paper, Metal, and Glass," *Washington Post,* September 23, 2014, accessed May 12, 2015, http://www.washingtonpost.com/blogs/wonkblog/wp/2014/09/23/americans-throw-out-more-food-than-plastic-paper-metal-or-glass/.

Big Ag takes bigger water: Amanda Kimble Evans, "Organic Methods Hold Water," Rodale's Organic Life, April 12, 2011, accessed April 27, 2015, http://www.rodalesorganiclife.com/home/organic-methods-hold -water.

Seventeen hundred gallons of water for one gallon of fuel: "Ethanol's Water Shortage," *Wall Street Journal*, October 17, 2007, accessed April 12, 2015, http://www.wsj.com/articles/SB119258870811261613.

Delaware strains under a 200-million-plus broiler bird industry: "Delaware's Growing Poultry Industry," Newsworks, August 11, 2014, accessed April 27, 2015, http://www.newsworks.org/index.php/ local/delaware/71394-delawares-growing-poultry-industry.

CHAPTER 16

According to the 2012 USDA census, there are nearly 24 percent more farms following this model since 2002: "2012 Census Drilldown: Organic and Local Food," National Sustainable Agriculture Coalition, May 2012, accessed April 12, 2015, http://sustainableagri culture.net/blog/2012-census-organic-local/.

the mightiest 2.2 percent of these vast operations control a full third of the nation's available acreage: "The Decline of the Small American Family Farm in One Chart," *Washington Post*, September 16, 2014, accessed February 15, 2015, http://www.washingtonpost .com/blogs/wonkblog/wp/2014/09/16/the-decline-of-the-small -american-family-farm-in-one-chart/.

According to USDA research, outside work accounted for 87 percent of American farmers' median income: "Table on Principal Farm Operator Household Finances, by ERS Farm Typology 2013," USDA Agricultural Resource Management Survey, 2013, accessed June 7, 2015, http://www.ers.usda.gov/topics/farm-economy/farm-household -well-being/farm-household-income-%28historical%29.aspx.

This magical new term encompasses farms grossing between $100,000 and $250,000: "Why Worry about the Agriculture of the

Middle?" University of Wisconsin, Madison, January 2012, accessed February 14, 2015, http://www.agofthemiddle.org/papers/whitepap er2.pdf.

Between 1997 and 2012 the number of these not-too-big, not-too-small types of operations declined by 18 percent: "MetLife Agricultural Investments: Ag Quarterly Newsletter," MetLife, accessed April 12, 2015, https://www.metlife.com/assets/cao/investments/Met LifeAgQuarterlySummer-2014.pdf, 2.

ten million Americans living in rural poverty: "Rural Research Brief: Poverty in Rural America," Housing Assistance Council, June 2012, accessed February 19, 2015, http://www.ruralhome.org/storage/re search_notes/rrn_poverty.pdf.

the loss of midsized farms "threatens to hollow out many regions of rural America": "Why Worry about the Agriculture of the Middle?"

CHAPTER 17

This mass production of meat chickens has been hailed as the most complete example of vertical integration in agriculture: "The Business of Broilers," Pew Charitable Trusts, January 1, 2013, accessed April 27, 2015, http://www.pewtrusts.org/~/media/legacy/uploaded-files/peg/publications/report/BusinessofBroilersReportThePew-CharitableTrustspdf.pdf.

Improper management of this broiler litter has led to polluted waterways and federal cleanups: "Big Chicken: Pollution and Industrial Poultry Production in America," Pew Environment Group, July 26, 2011, accessed April 28, 2015, http://www.pewenvironment.org /news-room/reports/big-chicken-pollution-and-industrial -poultry-production-in-america-85899361375.

There's more than one documented case where farmers have broken the cycle of servitude by suicide: Dave Murphy, "Farmers Look for Justice in the Poultry Industry," Cooking Up a Story, June 2, 2010, accessed April 28, 2015, http://cookingupastory.com/farmers-look-for

-justice-in-the-poultry-industry-met-with-fear-threats-intimidation
-and-hope-in-alabama.

nine out of ten farm households in the United States require some infusion of off-farm cash: Brett Wessler, "9 out of 10 Farm Households Collecting Off-farm Income," Drovers Cattle Network, September 13, 2013, accessed May 19, 2015, http://www.cattlenet work.com/news/industry/9-out-10-farm-households-collecting -farm-income.

CHAPTER 19

"Will the last one leaving Duluth please turn out the light?": Mike Creger, "Former Mayor Confirms Existence of Duluth's Fabled 'Turn Out the Light Billboard,'" *Duluth News Tribune*, September 30, 2014, accessed April 27, 2015, http://www.duluthnewstribune.com/content /former-mayor-confirms-existence-duluths-fabled-turn -out-light-billboard.

CHAPTER 20

The 2000 census reported a per capita income of just over thirteen thousand dollars: "Minnesota 2000: Population and Housing Units Counts," U.S. Census Bureau, accessed April 2013, https://www.census .gov/prod/cen2000/phc-3-25.pdf.

the tome of a book *The Amish* explains that both sects come from the Anabaptist tradition of sixteenth-century Europe: Donald Kraybill, Karen Johnson-Weiner, and Steven Nolt, *The Amish* (Baltimore: Johns Hopkins University Press, 2013), 4–9.

With growth like that, a community outgrows itself every twenty-five days, sending forth representatives to buy more land and start anew: "The Rural Sociologist," Rural Sociology, September 1, 2012, accessed April 28, 2015, http://www.ruralsociology.org/wp -content/uploads/2012/02/TRS-32-3.pdf, 34.

CHAPTER 21

the First Law of Thermodynamics tent, a crowded and popular place given that its tenants are the underpinning of modern neoclassical theories of economics: Cutler J. Cleveland, *The Economics of Nature and the Nature of Economics* (Cheltenham, UK: Edward Elgar, 2001), 16–17, 238.

The idea is that resources, even natural ones, enjoy exchangeability, and if you can't swap it out, there's likely a technofix: Barry W. Brook and Corey J. Bradshaw, "Strange Bedfellows? Techno-fixes to Solve the Big Conservation Issues in Southern Asia," *Biological Conservation Journal* 151 (2012): 7–10.

"The world can, in effect, get along without natural resources so exhaustion is just an event, not a catastrophe": John Bellamy Foster, "Ecology against Capitalism," *Monthly Review*, October 1, 2001, accessed April 27, 2015, http://monthlyreview.org/2001/10/01/eco logy-against-capitalism/.

Solow's circular First Law economic models to thinking of an animal only in terms of its circulatory systems, sans output: Herman E. Daly, *Steady-State Economics: Second Edition with New Essays* (Washington, DC: Island Press, 1992), 241.

CHAPTER 22

His data followed the nation's shift away from the early-twentieth-century model: Dr. Stewart Smith, "Sustainable Agriculture and Public Policy," *Maine Policy Review* 2, no. 1 (1993): 68–78, http://digi talcommons.library.umaine.edu/mpr/vol2/iss1/13.

Every action, from washing to peeling to dehydrating to the spraying on of sodium acid pyrophosphate . . . was now handled by the marketing sector: "Are You Eating Fake Mashed Potatoes?" Fooducate, accessed April 27, 2015, http://blog.fooducate.com/2010 /04/12/are-you-eating-fake-mashed-potatoes/.

Researcher Willard Cochrane calls it the Agricultural Treadmill: Virgil W. Dean, *An Opportunity Lost: The Truman Administration and the Farm Policy Debate* (Columbia: University of Missouri Press, 2006), 235.

CHAPTER 23

Mortality rates in outdoor laying hens dropped from 40 percent in the early 1900s to a mere 5 percent in modern caged operations: "American Egg Farming: How We Produce an Abundance of Affordable, Safe Food and How Animal Activists May Limit Our Ability to Feed Our Nation and World," United Egg Producers, http://www .unitedegg.org/information/pdf/American_Egg_Farming.pdf.

"not properly lighted as pullets": "Lighting for Small Scale Flocks," University of Maine Cooperative Extension Publications, accessed April 13, 2015, http://umaine.edu/publications/2227e/.

Poultry hysteria is a terrible condition: Dr. Ashleigh Bright, "Investigating Losses from Smothering in Commercial Poultry Flocks," Poultry Site, October 1, 2011, accessed April 4, 2015, http://www.thepoultrysite .com/articles/2194/investigating-losses-from-smothering-in -commercial-poultry-flocks/.

Index